Bayesian Computation Using Minitab

James H. Albert

Bowling Green State University

Duxbury Press

An Imprint of Wadsworth Publishing Company

I(T)P® An International Thomson Publishing Company

Belmont • Albany • Bonn • Boston • Cincinnati • Detroit • London • Madrid • Melbourne
Mexico City • New York • Paris • San Francisco • Singapore • Tokyo • Toronto • Washington

Editor: Alexander Kugushev
Editorial Assistant: Martha O'Connor
Production Editor: Jennie Redwtiz
Print Buyer: Barbara Britton
Permissions Editor: Peggy Meehan
Copy Editor: Laura Larson
Cover: Craig Hanson
Printer: Malloy Lithographing, Inc.

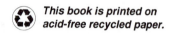 **This book is printed on acid-free recycled paper.**

For more information, contact Duxbury Press at Wadsworth Publishing Company.

Wadsworth Publishing Company
10 Davis Drive
Belmont, California 94002, USA

International Thomson Editores
Campos Eliseos 385, Piso 7
Col. Polanco
11560 México D.F. México

International Thomson Publishing Europe
Berkshire House 168-173
High Holborn
London, WC1V 7AA, England

International Thomson Publishing GmbH
Königswinterer Strasse 418
53227 Bonn, Germany

Thomas Nelson Australia
102 Dodds Street
South Melbourne 3205
Victoria, Australia

International Thomson Publishing Asia
221 Henderson Road
#05-10 Henderson Building
Singapore 0315

Nelson Canada
1120 Birchmount Road
Scarborough, Ontario
Canada M1K 5G4

International Thomson Publishing Japan
Hirakawacho Kyowa Building, 3F
2-2-1 Hirakawacho
Chiyoda-ku, Tokyo 102, Japan

Library of Congress Cataloging-in-Publication Data

Albert, James H.
    Bayesian computation using Minitab/James H. Albert.
        p.   cm.
    Includes index.
    ISBN 0-534-51781-1 (pbk.)
    1. Bayesian statsticcal decision theory—Data processing.
2. Minitab. I. Title.
QA279.5.A43   1997   96-1598
519.5'42—dc20

Contents

Preface

Why Was This Book Written?

This book describes the use of a package of Minitab programs for implementing Bayesian methods for a wide range of elementary statistical inference problems. The motivation for this project was my interest in using Bayesian ideas in teaching statistical inference at an elementary level. I have had difficulties in teaching the frequentist notion of inference and believe that the basic tenets of statistical inference may be better communicated using the Bayesian paradigm. I was encouraged to develop Bayesian teaching materials with the publication of Don Berry's text *Statistics: A Bayesian Perspective*. This book presents the basic inferential methods for proportions and means using Bayes' rule with a discussion of many real-life applications of these methods. One aspect of Berry's text that I found attractive is its focus on discrete models. In teaching Bayes' rule, I believe it is essential that the student get some experience specifying subjective probabilities, and prior probability distributions can be easier to assess when there are only a few values of the parameter of interest.

Although Bayes' rule is simple conceptually, it can be cumbersome to implement. For example, it can be tedious and time-consuming to compute posterior probabilities for two proportions when there are a large number of discrete models. Without the use of the computer, the instructor is forced in this situation to only discuss very simple models where only a couple of values of each proportion are used. This setting is not very realistic and does not motivate the more general setting where continuous models for the proportions are used. Thus, a general goal of these Minitab programs is to remove the computational burden in implementing Bayes' rule.

What Topics Are Covered in This Book?

This text describes and illustrates computer programs to perform Bayesian calculations for the inference topics in the standard one-semester introductory statistics course. Chapter 3 uses a program 'bayes' to introduce Bayesian inference when the unknown model is categorical. Chapters 4 through 8 describe the use of Minitab programs to address basic inference problems for one and two proportions, one and two normal means, and simple linear regression and contingency tables. These programs should help the instructor focus on the conceptual aspects of Bayesian inference. For example, in the case of learning about a proportion with discrete models, the instructor can focus his or her discussion on the specification of prior information and the summarization and interpretation of the posterior distribution rather than on the computation details.

Bayes' rule provides a simple mechanism to help us learn from data, and Bayesian methods can be used in a broad range of applications. In addition to the basic inference topics already described, this text also provides general algorithms and corresponding Minitab programs that can be used to implement Bayes' rule. Chapter 9 gives a general program for learning about a parameter when the prior is concentrated on a finite collection of values. This program is used to illustrate inference problems outside the realm of the usual binomial and normal problems. In particular, inference about Poisson, exponential, uniform, and hypergeometric populations is illustrated.

Simulation methods are demonstrated in this text for a range of problems. In Chapter 4, simulation is used to update probabilities about a proportion when the prior distribution is in the form of a histogram. The sampling-importance-resampling (sir) algorithm is used in Chapter 5 in estimating two proportions when the proportions are believed exchangeable. Chapters 10 and 11 implement simple algorithms for simulating from arbitrary prior/likelihood combinations. One nice feature of simulation methods is that a simulated sample of values from the posterior distribution is obtained, and data analysis techniques can be used to display and summarize this simulated sample.

The Organization of the Text

The Minitab programs are organized in the chapters by the inferential model. So all of the programs related to learning about a population proportion are contained in Chapter 4. The programs can also be divided by the type of inference. The *inference* programs focus on the computation of posterior probabilities or densities for the parameter or parameters of interest. The *predictive* programs compute probabilities about future observations. For example, the program 'p_disc_p' will compute predictive probabilities for a future binomial sample for a discrete probability distribution on a proportion. The *testing* programs focus on testing a particular statistical hypothesis by the use of a Bayes factor. The *prior assessment* programs are helpful in choosing a prior density to match one's beliefs about a parameter. The tables give a listing of the Minitab programs for each type of inference. For a particular type of inference, the table gives for each program the parameter(s) of interest and the method of representing prior opinion.

Minitab Inference Programs	
Program	Inference on
bayes_se, bayes	Categorical models using Bayes' rule
p_disc	One proportion using a discrete prior
p_beta	One proportion using a beta prior
p_hist_p	One proportion using a histogram prior
pp_disc	Two proportions using a discrete uniform prior
pp_discm	Two proportions using an arbitrary discrete prior
pp_beta	Two proportions using beta priors
pp_exch	Two proportions using an exchangeable prior
m_disc	One mean using a discrete prior
m_cont	One mean using a normal prior
m_nchi	One mean using a noninformative prior
mm_cont	Two means using normal distributions
mm_tt	Two means using t distributions
lin_reg	Linear regression
mod_disc	Arbitrary parameter using a discrete prior
mod_cont	Arbitrary parameter using a continuous prior

Minitab Prediction Programs	
Program	Prediction on
p_disc_p	Future binomial experiment using a discrete prior
p_beta_p	Future binomial experiment using a beta prior
m_cont	Future normal observation using a normal prior
lin_reg	Future response

Minitab Testing Programs	
Program	Testing on
p_beta_t	One proportion using a beta prior
pp_disct	Two proportions using a discrete prior
pp_bet_t	Two proportions using beta priors
m_cont_t	One mean using a normal prior
c_table	Independence in a contingency table
mod_crit	Two priors for discrete models

Minitab Prior Assessment Programs	
Program	Selecting a
beta_sel	Beta prior for a proportion
normal_s	Normal prior for a mean

How Can This Book Be Used?

This book can be used in a number of ways. It would be a suitable companion software text for instructors who plan to introduce some Bayesian inferential methods at an elementary level. It would also be suitable for instructors who wish to introduce Bayesian ideas and computing briefly in a statistical inference class for undergraduate or graduate students. The chapters of the book are independently presented, so an instructor can select out particular chapters and programs that are of interest. An instructor teaching probability may enjoy the programs that simulate games of chance in Chapter 2. The programs that implement basic summarization methods in Chapter 11 may be useful for a graduate class studying Bayesian computation.

Each chapter presents a particular inference method or computing strategy. After outlining the method, the computer programs are illustrated for a particular example. Further illustrations of the use of the programs are contained in the exercises. Because of space limitations, the book provides a limited explanation of Bayesian inference. See Berry[1], Antleman[2], or Lee[3] for basic expositions of Bayesian inference. However, it should be easy for an instructor to understand the methods and software after some exposure with Bayesian methods.

[1]Berry, D. (1996), *Statistics: A Bayesian Perspective*, Belmont, CA.: Duxbury Press.

[2]Antleman, G. (1996), *Elementary Bayesian Statistics*, Cheltenham: Edward Elgar Publishing.

[3]Lee, P. M. (1989), *Bayesian Statistics: An Introduction*, Oxford: Oxford University Press.

Why Minitab?

I decided to use Minitab since it is the most popular statistics package in teaching statistics. Many instructors are familiar with the basic syntax of Minitab, so it will not be difficult for them to learn how to use these new Bayesian commands. The programs are written using the "exec" type of Minitab macros and the higher resolution graphics that are available in Release 10 of Minitab and the Student Edition of Minitab for Windows[4]. Included in this disk is a second version of the programs that use the older style of character graphics. These programs work essentially the same as the ones described in the text and will run on Release 7 or later of Minitab.

Comparing Bayesian and Classical Procedures

The Bayesian procedures illustrated in this book are similar in some ways with classical statistical procedures. We briefly discuss the similarities and differences between classical and Bayesian methods for a problem of learning about a population proportion. For a particular Big Ten university, we are interested in estimating the proportion p of athletes who graduate within six years. For a particular year, forty-five of seventy-four athletes admitted to the university graduate. Assuming that this sample is representative of athletes admitted during other years, what have we learned about the proportion of all athletes who will graduate within six years? Specifically, we will consider two types of inferences. First, we wish to construct an interval that we are pretty confident contains the unknown value of p. Second, suppose that the university would like to state that over half of its athletes graduate on time — that is, p is larger than .5. Can the university make this statement with some confidence?

The classical 95% interval estimate for the proportion p for a large sample is given by

$$(\hat{p} - 1.96\sqrt{\frac{\hat{p}(1 - \hat{p})}{n}}, \hat{p} + 1.96\sqrt{\frac{\hat{p}(1 - \hat{p})}{n}})$$

where $\hat{p}$ denotes the sample proportion of athletes who graduate within six years, and

---

[4]McKenzie, J., Schaefer, R. L., Farber, E. (1995), *The Student Edition of MINITAB for Windows*, Reading, MA.: Addison-Wesley.

n is the size of the sample. In this example $\hat{p} = \frac{45}{68} = .608$ and $n = 68$ and, by substitution in the above formula, one obtains the interval $(.497, .719)$.

If the university wishes to show that the proportion p is larger than one half, then the classical approach would test the null hypothesis H: $p \leq .5$ against the alternative hypothesis K: $p > .5$. One decides between the hypotheses on the basis of a p value. In this example, one finds the probability that the sample proportion $\hat{p}$, in repeated sampling, is equal to or greater than the observed value $\hat{p} = .608$ when $p = .5$. Using binomial tables, one finds that the p value is equal to .0403. If the level of significance is .05, one concludes that there is sufficient evidence that the proportion of athletes who graduate exceeds one-half.

The Bayesian approach to learning is based on the subjective interpretation of probability. The value of the proportion p is unknown, and a person expresses his or her opinion about the uncertainty in the proportion by means of a probability distribution placed on a set of possible values of p. The *prior distribution* is the probability distribution that the person has before observing data. After observing data, the person changes his or her opinion about the value of the proportion. The new probability distribution, the *posterior distribution*, is computed using Bayes' rule. All of the person's knowledge about the proportion is contained in the posterior distribution, and statistical inferences are made by summarizing this distribution.

Let us reanalyze our example from a Bayesian perspective. Suppose that little is known about the location of the proportion p. We construct a prior distribution for this proportion that reflects this belief. Suppose that p could conceivably be any one of the ninety-nine values .01, .02, ..., .99. If we know very little about the proportion, we may think that these values for p are equally likely, and so we assign to each value the probability 1/99.

After observing the graduation results, we update our probability distribution for p using Bayes' rule. The Minitab program 'p_disc', discussed in Chapter 4, can be used to compute the posterior probabilities. The values of the proportion and the corresponding probabilities are placed into two columns of the Minitab worksheet. When the program is run, we input the number of football players who graduate (forty-five) and the number who do not graduate (twenty-three). The posterior probabilities are computed by the program and placed in a new column of the worksheet.

The posterior probabilities represent our current opinion about the graduating proportion of the football players. A 95% probability interval for the proportion is found by finding a collection of p values with a probability content that is approximately .95. The Minitab program 'disc_sum' is designed to find such a probability interval. Using this program, we obtain the set $\{.50, .51, \ldots, .71\}$, which is approximately equal to the 95% confidence interval found using the frequentist method.

Although the classical and Bayesian intervals agree, the interpretations of the two intervals are different. The probability of the proportion p falling in the Bayesian interval [.50, .71] *for this data set* is actually 95%. In contrast, from a classical perspective, one is not confident that the interval (.497, .719) contains p. The classical statistician is confident about the *procedure* — if this 95% interval is computed for repeated sampling, then we are confident that approximately 95% of the intervals will contain the proportion value.

Next, consider the question whether over half of the athletes graduate within six years. From a Bayesian viewpoint, the plausibility of the hypothesis $H : p \leq .5$ is found by computing its posterior probability. From the set of posterior probabilities, one finds that the probability the proportion value is less than or equal to one-half is .032. This probability is small, so we would conclude that there is good evidence that over half of the athletes graduate on time. Note that the posterior probability of the hypothesis H is approximately equal to the classical p value.

For this example, the classical procedures gave similar answers to the Bayesian procedures when a weak or noninformative prior distribution was used. However, there are important distinctions between the two sets of procedures. One difference is the interpretation. The Bayesian computes probabilities about the unknown proportion p conditional on the sample that is observed. In contrast, the classical statistician has no confidence that he of she is correct for this data set. The confidence comes from repeating this inferential process for many data sets.

The Bayesian mode of inference has a number of desirable features. One attractive feature is that inferential statements about a parameter are easy to communicate. It is natural to talk about the probability that p falls in an interval or the probability that a hypothesis is true. A second nice feature is that a single recipe, Bayes' rule, is used for updating one's probabilities about a parameter. This rule can be used for small or large sample sizes. Similarly, there is a straightforward Bayesian recipe for predicting the values of observations in a future. In this example, suppose that there is a new sample of 100 athletes from the university. The Minitab program 'p_disc_p' can be used to predict the number of athletes that will graduate in this sample. Last, Bayesian methods allow a person to use his or her subjective beliefs about the location of the parameter in the inference problem. In our example, one may have some opinions about graduating rates of athletes based on data from other universities, and a prior probability distribution for p can be constructed to reflect this knowledge. Bayes' rule provides a useful mechanism for combining this prior knowledge about the graduation rates with information contained in the sample.

Acknowledgments

I am appreciative to Don Berry for his support and encouragement during the development of this software. Much of this work was done during a sabbatical leave at the Institute of Statistics and Decision Sciences at Duke University, and I am thankful for the warm support of the ISDS faculty during this time. The Minitab programs were tested in a statistics course at Duke, and Jon Stroud was very helpful in assisting me in this course. I am grateful for the many helpful suggestions provided by the reviewers: Asit P. Basu, University of Missouri–Columbia; Michael Brimacombe, University of Pittsburgh; Cindy L. Christiansen, Harvard Medical School; Mauro Gasparini, Purdue University; and H. A. Germer, Mississippi County Community College. In addition, Lisbeth Cordani, Dave Dole, Telba Irony, Mary Kay Porter, Allan Seheult, and Patricia Williamson provided useful comments in the development of this work. Jim Berger has been a great influence in my training in statistics. Last, but certainly not least, I thank my wife, Anne, and children, Lynne, Bethany, and Steven, for their understanding and patience during the completion of this project.

Chapter 1

An Introduction to Minitab

1.1   Overview

Minitab is a system for performing statistical calculations suitable for a wide range
of statistical analyses. Moreover, it is relatively easy to learn and the most popular
program used in teaching concepts of probability and statistics. Our purpose in this
chapter is to introduce some basic Minitab commands that will be helpful in using the
programs described in this book.

Minitab performs calculations on numbers that are stored in an electronic worksheet.
In Figure 1.1, the opening Minitab window for the Student Edition of Minitab for
Windows is displayed. The Data window shows the worksheet; columns of this worksheet
are labeled by C1, C2, and so forth. In addition, numbers can be placed into stored
constants called K1, K2, and so forth. In Figure 1.1, note that there is a second large
window called the Session window. Calculations can be performed by typing commands
in the Session window. All of the Minitab text output will appear in this window.
Commands can also be executed by the use of menus; however, since the programs of
this book are not available using menus, we will focus on the use of commands entered
into the Session window.

1.2   Entering data

We will get started by entering numbers in the worksheet. Suppose that you are interested in summarizing a small discrete probability distribution. Call the variable p. It

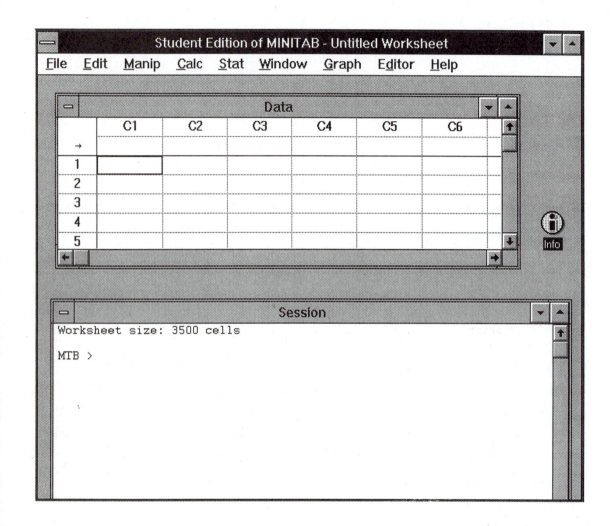

Figure 1.1: Opening screen of the Student Version of Minitab for Windows

can take on the five values .1, .2, .3, .4, .5, with respective probabilities .1, .2, .4, .2, .1. The easiest way to enter these numbers in the worksheet is to type numbers directly into the Data window. To name the first column 'p', place the mouse in the entry field or square immediately under the C1 label. Type "p" in this cell. Then the five values for 'p' can be typed in the rows 1, 2, 3, 4, 5 under this label. Likewise, column C2 can be named 'prior' and the five probabilities are placed in the first five rows in this second column.

An equivalent way to enter data in the worksheet is to type commands in the Session window after the MTB prompt. This method will be illustrated here since all special Minitab programs (macros) are also executed in the Session window. In the output below, we first give names to columns C1 and C2 of the worksheet using the "name" command. Next, we enter the previous numbers into the two columns using two "set" commands.

```
MTB > name c1 'p' c2 'prior'
MTB > set 'p'
DATA> .1 .2 .3 .4 .5
DATA> end
MTB > set 'prior'
DATA> .1 .2 .4 .2 .1
DATA> end
```

To indicate that we have stopped entering numbers into a column, we used the "end" command. By the way, column names and commands in Minitab are not case-sensitive. The last command in the earlier output could have been typed

```
SET 'PRIOR'
```

To see the current contents of the worksheet, the "info" command is used. This command, shown here, confirms what we have done using the name and set commands. The first column C1 is named 'p'; the "Count" heading indicates that there are five elements in this column.

```
MTB > info
Column    Name        Count
C1        p             5
C2        prior         5
```

1.3   Computations on the worksheet

Once numbers have been entered in the Minitab worksheet, many commands are available to display and perform computations on the columns in the worksheet. To display the numbers in a set of columns, use the print command. To verify that the two columns have been entered correctly, display the contents of columns 'p' and 'prior':

```
MTB > print 'p' 'prior'
 ROW       p  prior
   1     0.1    0.1
   2     0.2    0.2
   3     0.3    0.4
   4     0.4    0.2
   5     0.5    0.1
```

To check that we have indeed entered a proper probability distribution, use the "sum" command:

```
MTB > sum 'prior'
   SUM     =        1.0000
```

The printed output is the sum of the entries in the column 'prob'; the value is 1, so a legitimate set of probabilities has been entered.

We will next compute the mean of the probability distribution and will do this in two steps. First, we name a new column 'product'. The let command shown will enter five new values in the column 'product'. The entry in a particular row in 'product' will be the product of the values of 'p' and 'prior' in the same row. Then the sum command is used to find the sum of the values in 'product'. The printed value, .3, is the mean M_1 of the probability distribution.

```
MTB > name c3 'product'
MTB > let 'product'='p'*'prior'
MTB > sum 'product'
   SUM     =        0.30000
```

You can perform this computation without defining a new column. The let command below first multiplies the individual elements of the columns 'p' and 'prior', sums the products, and then stores the answer in the Minitab constant K1. The print command is used to display the contents of this constant.

```
MTB > let k1=sum('p'*'prior')
MTB > print k1
K1       0.300000
```

To compute the standard deviation of the distribution, you first must square each of the values of 'p', multiply each squared number by its probability, and then find the sum of the products to find the second moment M_2:

```
MTB > let k2=sum('p'**2*'prior)
```

Next, the standard deviation is found using the formula

$$S = \sqrt{M_2 - M_1^2}.$$

In Minitab, the syntax is

```
MTB > let k3=sqrt(k2-k1**2)
MTB > print k3
K3       0.109544
```

The displayed value of the constant K3 is the value of the standard deviation of the posterior distribution.

To make probability computations, a number of special Minitab commands are useful. To find cumulative probabilities, you can define a new column 'cumprob' and place partial sums of the column 'prior' into this new column by the "parsums" command.

```
MTB > name c4 'cumprob'
MTB > parsums 'prior' 'cumprob'
MTB > print 'p' 'cumprob'
 ROW      p  cumprob
   1    0.1      0.1
   2    0.2      0.3
   3    0.3      0.7
   4    0.4      0.9
   5    0.5      1.0
```

In this printout, the contents of 'p' and 'cumprob' are shown. We see that the probability that 'p' is less than or equal to .3 is .7.

The let command can also be used with logical operators to find more complicated probabilities. Suppose that we wish to find the probability that 'p' is between .2 and .4 inclusive. We use these commands displayed:

```
MTB > let k2=sum('prior'*('p'<=.4)*('p'>=.2))
MTB > print k2
K2        0.800000
```

The expression ('p'<=.4) creates a column of 0's and 1's. A row value is equal to 1 if the value in 'p' is less than or equal to .4 and 0 otherwise. Here the column would contain the values 1, 1, 1, 1, 0. Likewise, the expression ('p'>=.2) would contain the values 0, 1, 1, 1, 1. If these two columns and the column 'prior' are multiplied, then the resulting column contains the values 0, .2, .4, .2, .0. The sum command will add up the numbers of this last column, and the value (the probability of interest) is stored in the constant K2. The print command displays the value of this constant.

1.4   Descriptive statistics and graphs

To illustrate some Minitab commands for summarizing and graphing data, consider a simple probability simulation. We take a simulated sample from a beta density with parameters 2 8 by typing the commands

```
MTB > rand 500 'p';
SUBC> beta 2 8.
```

The basic command is "rand"; this command will simulate 500 random variates and place them in the column named 'p'. This command illustrates the use of Minitab subcommands. By putting a semicolon at the end of the line, we are able to enter the subcommand "beta 2 8" (ending with a period). This indicates that the distribution we wish to simulate from is beta with parameters 2 and 8. Other subcommands used with rand allow simulation from other probability distributions such as the normal and uniform.

Many Minitab commands are available for summarizing a batch of numbers. To get basic descriptive statistics (mean, standard deviation, etc.) for the simulated data in the column 'p', we type

```
MTB > describe 'p'
             N      MEAN    MEDIAN    TRMEAN     STDEV    SEMEAN
p          500   0.19434   0.18710   0.18807   0.11484   0.00514

           MIN       MAX        Q1        Q3
p      0.00780   0.55335   0.10007   0.25875
```

In the output, we note that the mean of the simulated values of .194 and 50% of the simulated values fall between the first and third quartiles, .100 and .258.

Minitab has many types of graphs available for displaying numbers. To get a dotplot (a simple graphical display) of the data, we type

```
MTB > dotplot 'p'
```

```
                            .    :
                       .    :   .:
                  .         :  :  ::
               ::: :       :: : .::
             : :::.: .     :: :::::
             ::::::::::     :: :::::      :
             .::::::::::    ::.:::::::.:  :
            :::::::::::::..::::::::::::: :::
            ::::::::::::::::::::::::::::::: . .
           :::::::::::::::::::::::::::::::: :.:  . .     .
           ::::::::::::::::::::::::::::::::::.:::..:::   : ..
           ::::::::::::::::::::::::::::::::::::::::::::.:::::.: ...:
         +---------+---------+---------+---------+---------+-------p
        0.00      0.10      0.20      0.30      0.40      0.50
```

The dotplot gives a picture of the beta$(2,8)$ probability density. We see that the density is mound-shaped with all of the values falling between 0 and .6.

Another useful display is a histogram. We can graph a histogram of our simulated data by typing

```
MTB > hist 'p'
```

The Windows version of Minitab displays this figure in a separate Graph window; Figure 1.2 shows the histogram. This plot gives essentially the same picture of the data as the earlier dotplot.

1.5 Saving and printing your work

To save the contents of the session command to a file, the Minitab "outfile" command is useful. To start saving the output to a text file named 'myfile', type in the Session window

```
outfile 'myfile'
```

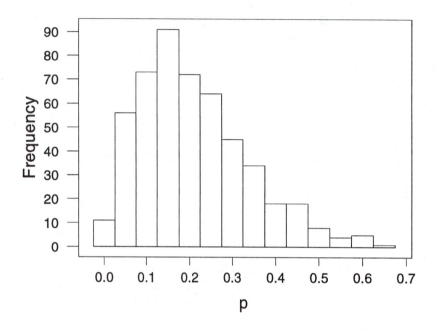

Figure 1.2: A histogram of a simulated sample from a beta(2, 8) distribution.

When you want to stop storing output, type

`nooutfile`

Printing the contents of the Session window and Graph windows can be done using the Print option in the File menu. Suppose that you wish to print a particular graph. Click on this window to make it active and then use the print menu item.

1.6   Using the macros

All of the programs described in this book are written using Minitab "exec" macros. A *macro* is a collection of Minitab commands stored as a text file of the form 'name.MTB', where 'name' is the name of the particular program. Before these programs can be run, you have to tell Minitab where the collection of macro files is located on your computer.

One can use these files from the disk included in this book. On this disk, the programs are stored in the subdirectory "bayes." It may be faster to access these programs if they are stored on your hard drive. In this case, it is recommended that all of the programs (each ending with the suffix MTB) be copied to a subdirectory named "bayes" on the hard drive. After Minitab has been started, make this subdirectory (the one on the floppy disk or the hard drive) the local directory on Minitab. If the hard drive is denoted "c," then make this subdirectory the local directory by typing in the Session window the command

```
cd c:\bayes
```

All of the programs work in a similar manner. First, it may be necessary first to name and place data into particular columns of the worksheet. For example, the macro 'p_disc' requires that there exist two columns named 'p' and 'prior', and numbers should be placed in each of these two columns. These columns can be created directly in the Data window or by using name and set commands in the Session window. Then to run a program, say 'p_disc', you type in the Session window the command

```
exec 'p_disc'
```

This command must be typed exactly as shown, or an error will result. In the running of the program, different information will be requested. The program 'p_disc' asks for the number of successes and failures in the data set. Then output from the program is displayed in the Session window. Some graphs may be produced, which will appear in separate Graph windows.

These programs perform operations on the Minitab work space. Generally, they work using the remote columns C51–C100 to avoid any conflict with columns that the user may have previously defined. However, it can happen that, in the running of these programs, there will be a conflict between names of columns already existing and names that the program will want to define. If there is a conflict in column names and an error does result, one simple cure is initially to erase all of the Minitab columns by typing

```
erase c1-c100
```

and then running the program.

Many of the programs are named using the general convention

parameter_model_inference

So the program 'p_disc_p' is a program designed to learn about a proportion parameter p, *discrete* models are used, and the inference is *prediction*. Similarly, the program 'm_cont_t' concerns a population mean M using *continuous* models and a *test* is the inference.

Chapter 2

Simulating Games of Chance

2.1 Introduction

This chapter introduces basic notions of probability by using Minitab to simulate three popular games of chance. The Minitab program 'craps' simulates the popular casino game. One plays by repeatedly tossing two dice; the outcome of the game depends on the sum of the numbers of the two dice. By simulating this game a large number of times, one can observe the likelihoods of different outcomes. Craps provides a good illustration of the idea of conditional probability, which is central to understanding Bayes' rule. The program 'yahtzee' simulates a simplified version of the popular dice game Yahtzee. This game is played by tossing a set of five dice three times with the option of keeping desirable rolls at each turn. The object is to get particular patterns of dice, such as a full house or five of a kind. This program illustrates how one can approximate probabilities using relative frequencies. It also is useful for finding game strategies that maximize the probabilities of dice roll patterns of interest. The last program, 'bball', simulates a baseball season between a set of teams. One inputs numbers that reflect the strengths of the teams, and a baseball season is simulated using a paired comparison model. This model, although simple, gives patterns of wins and losses that can approximate pretty well the observed wins and losses of teams in major league baseball. This program illustrates the phenomenon that the "best" team (the team with the highest probability of winning) may not be the team with the best win-loss record at the end of the season.

2.2 Playing craps

> Minitab command to play craps:
> **exec 'craps'**

The basic version of the casino game craps is played as follows. The player rolls a pair
of dice and observes the sum of the numbers appearing on the two dice. If the sum of
the two dice is 2, 3, or 12, the player loses the game. If the sum of the rolls is 7 or 11
the player wins. If another number is thrown, then this number becomes the "point,"
and the play of the game continues. The player will roll the dice repeatedly until either
a sum of 7 or the point is thrown again. If the point appears first, the player wins, and
if a 7 appears first, the player loses.

A sample run of the Minitab program 'craps' is displayed here. Input the number
of games, and the program displays the dice rolls for each game and the final outcome
(win or loss). For game 1, the first sum of the first dice was 5. So in this game the sum
of 5 is the point, and one keeps rolling until a sum of 5 or a sum of 7 is observed. In
the fifth roll, a sum of 7 occurs and the player loses this particular game. Each time the
game is played, the program records three quantities. The column 'WIN?' records the
win or loss — win is recorded as a 1 and a loss as a 0 (in this particular game, 'WIN?'
= 0). The column '1ST_ROLL' records the sum of the two dice on the first roll. In the
first game, this sum is 5. The column 'N_ROLL' records the total number of dice rolls
made; in the first game 'N_ROLL' = 5.

```
MTB > exec 'craps'
HOW MANY GAMES DO YOU WANT TO PLAY?
DATA> 2
NOTE:
  Type 'y' and return to play:
y
GAME
   1
ROLLS
    5    10     6     4     7
RESULT
  You lose!
NOTE:
  Type 'y' and return to play:
y
```

```
GAME
   2
ROLLS
   5    9    7
RESULT
  You lose!
```

At the end of the run of the program 'craps', tables of counts are displayed that categorize the wins and losses by the first roll and by the total number of dice rolls. If this game is played many times, then one can approximate various probabilities and conditional probabilities of interest. The program 'craps' was rerun, playing 100 games, and the two tables of counts summarizing these 100 games are displayed here.

The first table classifies the 100 games by whether you won or not (column 'WIN?') and the sum of dice on the first roll (column '1ST_ROLL'). How often did you win the game? We see that we won 44 of 100 games, so the estimated probability of winning is .44 = 44/100. Some patterns in these counts of the two-way table are obvious based on the rules of craps. For example, note that a sum of 7 was rolled first for sixteen games and every game of these sixteen (as expected) was a winner. But do some point values give higher chances of winning? For example, for the nine games that you first rolled a sum of 4, you only won two, for a winning percentage of 2/9=22%. In contrast, the value 6 was a point thirteen times, and you won five games for a winning percentage of 5/13=38%. This suggests that you have a better chance of winning with a point value of 6 than for a point value of 4. These conclusions should be made with some caution since we are only computing approximate probabilities based on relatively small sample sizes. One can better approximate these conditional probabilities by rerunning 'craps' and playing a larger number of games.

ROWS: 1ST_ROLL	COLUMNS: WIN?		
	0	1	ALL
2	8	0	8
4	7	2	9
5	13	5	18
6	8	5	13
7	0	16	16
8	7	4	11
9	9	4	13
10	4	1	5

```
11          0        7         7
ALL        56       44       100
```

```
ROWS: N_ROLLS      COLUMNS: WIN?
              0        1      ALL
   1          8       23       31
   2         14        6       20
   3         12        4       16
   4          7        4       11
   5          6        3        9
   6          3        0        3
   7          0        1        1
   8          0        1        1
   9          1        0        1
  10          1        1        2
  14          0        1        1
  16          1        0        1
  17          1        0        1
  18          1        0        1
  19          1        0        1
 ALL         56       44      100
```

The second table from the displayed output of the program categorizes the 100 games by 'WIN?' and the number of rolls (column 'N_ROLLS'). Did the game end frequently after one roll? Totaling the counts in the first row, we see that the game ended after the first roll thirty-one times for a probability of $31/100 = .31$. If the game drags on, do you have a better chance of winning? Note from the output that your winning fraction is $23/31 = .74$ for one-roll games and $21/69 = .30$ for games that lasted more that one roll. So it appears from this brief analysis that you have a much better chance of winning short games that last one roll than long games.

2.3 Playing Yahtzee

> Minitabs commands to play Yahtzee:
> Player game:
> **exec 'yahtzee'**
> Automatic game:
> **exec 'yahtz_au'**
> Automatic repeat game:
> **exec 'yahtz_re'**

The game that is simulated using the Minitab program is a simplified version of the dice game Yahtzee. One rolls a set of five dice three times to get particular patterns on the dice. We will consider nine possible patterns. The first eight will be described in increasing order of their value. A "one-pair" is a sequence of rolls such as 1, 2, 2, 3, 4, where exactly one number (here 2) occurs twice. A more valuable pattern, a "two-pair," is when two numbers each occur twice, such as 2, 3, 4, 3, 4. A "3-of-a-kind" is when exactly one number occurs three times, such as 5, 4, 4, 4, 6. A "full house" is when one number occurs twice and a second number occurs three times, such as 6, 6, 3, 3, 6. Also, sequences of consecutive die numbers are of interest. A "small straight" is when exactly four numbers occur in sequence such as 2, 3, 4, 5, 5. A "large straight" is when five numbers occur in sequence, such as 1, 2, 3, 4, 5. A "4-of-a-kind" is the pattern where one die number occurs four times such as 5, 5, 5, 3, 5 and a "yahtzee" is when the same number appears on all five dice. If none of these patterns occur, we record a "nothing" — this pattern is the least valuable.

A player wishes to obtain the most valuable pattern of dice using three rolls. After each roll, she has the option of saving particular dice and only rolling the remaining. One can play this game using the Minitab macro 'yahtzee'. The output of one game is displayed on p. 16. After the player first types 'y' to start the game, the program gives the sequence of five numbers of the first roll. In this game, the dice numbers are 6, 5, 3, 6, 1. The player enters a sequence of 0's and 1's that indicate the dice to keep. Here one has rolled two 6's; suppose that the person wishes to save these two dice and discard the remaining. This choice is made by inputting the numbers 1, 0, 0, 1, 0; this indicates that the person is saving dice 1 and 4 and discarding dice 2, 3, and 5. The program then rolls dice 2, 3 and 5 again — the next set of rolls 6, 4, 2, 6, 5 is output. Since this roll includes the consecutive numbers 4, 5, 6, it may be better at this point to save these three rolls with the hope of getting a small or large straight. So the player saves dice 2, 4, and 5. Dice 1 and 3 are rolled again in this last turn. The final roll is 3, 4, 6, 6, 5, which is a small straight.

```
MTB > exec 'yahtzee'
Type 'y' and return to start rolling dice:
y
rolls
   6    5    3    6    1
result
  1 pair
ENTER (USING SEQUENCES OF 0'S AND 1'S) WHICH DICE TO KEEP:
DATA> 1 0 0 1 0
rolls
   6    4    2    6    5
result
  1 pair
ENTER (USING SEQUENCES OF 0'S AND 1'S) WHICH DICE TO KEEP:
DATA> 0 1 0 1 1
rolls
   3    4    6    6    5
result
  small straight
```

After playing this game a number of times, one thinks about alternative strategies for discarding dice to obtain valuable patterns of dice. One particular strategy for playing Yahtzee is implemented using the Minitab program 'yahtz_au' (named for Yahtzee, automatic mode). The output of one run of this macro is displayed next. The program shows the dice rolls and automatically decides which dice to save at each turn (the saved dice are indicated by the label "keep"). The computer's strategy will be to save any die numbers that are repeated in a particular roll. The exception to this rule is that the numbers of a small or large straight will be kept. By running 'yahtz_au' a few times, one can understand the strategy that is used.

```
MTB > exec 'yahtz_au'
Type 'y' and return to start rolling dice:
y
rolls
   1    1    4    3    3
result
  2 pair
keep
   1    1    0    1    1
TYPE 'y' AND RETURN FOR NEXT ROLL:
y
```

```
rolls
   1    1    1    3    3
result
  full house
keep
   1    1    1    1    1
TYPE 'y' AND RETURN FOR NEXT ROLL:
y
rolls
   1    1    1    3    3
result
  full house
```

How does this computer strategy perform in the long run? After running the program 'yahtz_au' one time, one can use the macro 'yahtz_re' to run the game in the automatic mode a large number of times. In the output below, 'yahtz_re' is executed ninety-nine times. For each play of the game, the program keeps track of the pattern of the dice after the first, second, and third rolls. The pattern numbers of these three rolls are stored in the columns 'ROLL_1', 'ROLL_2', 'ROLL_3', respectively. (The pattern numbers of the nine possible outcomes are listed in the table.) In some cases, the game is completed at the end of the first or second roll. For example, if a large straight is obtained after the first roll, then rolls 2 and 3 are not necessary. In this case, 'ROLL_1' is set equal to 8, and 'ROLL_2' and 'ROLL_3' are both set equal to 0. The pattern number of the final roll is stored in the column 'F_ROLL'.

Pattern Number	Outcome
1	Nothing
2	1 pair
3	2 pair
4	3-of-a-kind
5	Full house
6	Small straight
7	Large straight
8	4-of-a-kind
9	Yahtzee

```
MTB > exec 'yahtz_re' 99

MTB > prin 'ROLL_1' 'ROLL_2' 'ROLL_3' 'F_ROLL'
 ROW   ROLL_1  ROLL_2  ROLL_3  F_ROLL

   1       3       5       0       5
   2       7       0       0       7
   3       3       3       3       3
   4       3       3       3       3
   5       2       8       8       8

  97       2       4       8       8
  98       2       6       7       7
  99       6       7       0       7
 100       3       3       3       3

MTB > table 'ROLL_1' 'F_ROLL'
 ROWS: ROLL_1      COLUMNS: F_ROLL
            3       4       5       6       7       8       9     ALL
   1        1       0       1       0       0       1       0       3
   2       15       4       7       2       1       8       0      37
   3       15       0       9       0       0       0       0      24
   4        0       2       3       0       0       7       0      12
   5        0       0       5       0       0       0       0       5
   6        0       0       0       4       3       0       0       7
   7        0       0       0       0       7       0       0       7
   8        0       0       0       0       0       4       1       5
 ALL       31       6      25       6      11      20       1     100
```

In the output, some of the results of the 100 "automatic" games are displayed by listing the contents of the columns 'ROLL_1', 'ROLL_2', 'ROLL_3', and 'F_ROLL'. By tabulating the values of these columns, one can learn about the likelihoods of various events and compute some conditional probabilities. In the output, the Minitab "table" command is used to classify the 100 games by the pattern number of the first roll and the pattern number of the final roll. How likely is it to get a Yahtzee in this computer strategy? A Yahtzee corresponds to pattern number 9, and the variable 'F_ROLL' was equal to 9 one time, indicating that the approximate probability of a Yahtzee is $1/100$ = .01. The probability of ending with a full house after three rolls is about $25/100$ = .25. If one obtains a 3-of-a-kind (pattern number 4) on the first roll and follows this computer strategy, then what is the chance of a 4-of-a-kind (pattern number 8) on the final roll? From this table, we see that this conditional probability P(4-of-a-kind on final roll given 3-of-a-kind on first roll) is high — approximately $7/12 = .58$. If one obtains a small straight on the first roll, the conditional probability of a large straight by the third roll is approximately $3/7 = .43$. Probability computations such as these may be helpful in designing alternative strategies for playing the game.

2.4   Playing baseball

Minitabs commands to simulate a baseball season:
To perform one simulation:
exec 'bball'
To simulate many seasons:
exec 'bball_re'

The program 'bball' simulates a baseball season played between a group of teams using a simple probability model. Suppose, for illustration, there are five teams in the baseball league; call the teams A (Atlanta), B (Boston), C (Cleveland), D (Detroit), F (Florida). Each team is assigned a number that reflects its strength. If equal numbers of teams have positive and negative strengths, a strength value of 0 corresponds to a team that will win approximately half of its games in the long run. Positive values of strength correspond to winning teams and negative values to losing teams. The probability that one team defeats a second team is determined by the strength values. Specifically, suppose that Atlanta plays Boston and the teams have strength values given by s_A and s_B, respectively. Then the probability that Atlanta wins the particular game is given by $p = \frac{\exp(s_A)}{\exp(s_A)+\exp(s_B)}$. A game is simulated by generating a random number u that is uniformly distributed from 0 to 1. If u is less than the probability p, Atlanta wins the game; otherwise Boston wins. This procedure is repeated for an entire season of games in which each team plays every other team eight times.

In the run of the program 'bball' displayed here, one inputs a set of team strengths. Here the numbers $-1, -.5, 0, .5, 1$ are used. Since five numbers are input, this indicates that there are five teams with the given strengths. The program simulates games between all five teams and outputs the results of the season using two tables. The first table gives the number of games each team wins against every other team. For example, we see that of the eight games played between teams 1 and 2, team 1 won 5 and team 2 won 3. The second table displays the total number of wins for all of the teams. Since teams 1, 2, 3, 4, 5 have increasing strength values, it is interesting to see whether the order of finish in the thirty-two game season corresponds to their abilities. In this particular season, team 5 was the pennant winner with a record of 24–8. The teams did not finish in the order of their abilities — team 2 with the second smallest ability had the poorest record of 8–24 and finished last.

```
MTB > exec 'bball'
INPUT TEAM STRENGTHS:
DATA> -1 -.5 0 .5 1
DATA> end
ROWS: winner      COLUMNS: loser
              1           2          3          4          5         ALL
   1          0           5          4          1          1          11
   2          3           0          1          2          2           8
   3          4           7          0          1          2          14
   4          7           6          7          0          3          23
   5          7           6          6          5          0          24
  ALL        21          24         18          9          8          80

  winner   COUNT
       1      11
       2       8
       3      14
       4      23
       5      24
      N=      80
```

After the program 'bball' is run once, one can simulate more baseball seasons using the macro 'bball_re'. This program will use the same set of strength values that were input in the program 'bball'. In the output here, additional seasons are simulated by the exec command where the number following the name of the program corresponds to the number of additional simulated seasons. As the program is run, the numbers of games won by all teams are output for each simulation. The games won by teams 1, 2, and so on, are stored in the columns 'n_win_1', 'n_win_2', ...; the place finishes by the teams are stored in the columns 'place_1', 'place_2',.... In the run of the program here, nine additional seasons are simulated, and the results of the ten (nine plus the first simulation) are displayed by printing columns 'n_win_1'–'n_win_5' and 'place_1'–'place_5'.

If one looks at the season performances for a particular team in the output, one notices substantial variability across seasons. For example, the number of wins by the best team (team 5), ranged from 19 to 26 for these ten seasons. Team 3, which has an average ability, finished first in season 3 and fourth in season 9. The general impression that one gets from this simulation is that it can be difficult to distinguish between the qualities of these teams on the basis of a thirty-two game season.

```
MTB > exec 'bball_re' 9
MTB > print 'nwins_1'-'nwins_5'

Row   nwins_1   nwins_2   nwins_3   nwins_4   nwins_5
  1        11         8        14        23        24
  2         9        11        17        17        26
  3         9        18        20        14        19
  4         4        15        17        18        26
  5         8        11        15        21        25
  6         6        13        15        22        24
  7        12         8        18        20        22
  8         5        16        18        18        23
  9         7        17        13        18        25
 10         7        10        17        21        25

MTB > print 'place_1'-'place_5'
Row   place_1   place_2   place_3   place_4   place_5
  1         4         5       3.0       2.0         1
  2         5         4       2.5       2.5         1
  3         5         3       1.0       4.0         2
  4         5         4       3.0       2.0         1
  5         5         4       3.0       2.0         1
  6         5         4       3.0       2.0         1
  7         4         5       3.0       2.0         1
  8         5         4       2.5       2.5         1
  9         5         3       4.0       2.0         1
 10         5         4       3.0       2.0         1
```

2.5   Exercises

1. Play craps on Minitab 200 times. Let's focus on the table in the output that classifies rolls by win or loss (column 'WIN?') and the number of the first roll ('1ST_ROLL'). Answer the following questions based on this table.

   (a) How many times did you win? Do you believe that this is a fair game?

   (b) What is the chance that you lose on the first roll of the game (i.e., roll a 2, 3, or 12)?

   (c) Suppose that you roll a 10 on the first roll. What is the chance that you will win?

   (d) For each of the eleven possible first rolls, find the probability that you win the game. Are there particular numbers that you can roll first (in addition to 7) such that you have a good chance of winning?

   (e) Look only at the games you won. Given that you won the game, what is the chance that you first rolled a 7?

2. Play Yahtzee in the automatic mode 100 times.

   (a) Using the command

      ```
      tally 'ROLL_1'
      ```

      find the frequencies of the nine possible outcomes on the first roll. What was the most likely outcome? What was the least likely? What is the probability of a "nothing" roll?

   (b) Using the tally command for the numbers in the column 'F_ROLL', summarize the 100 final rolls. Find the most likely outcome, the least likely outcome, and the probability that the final roll is "nothing." Compare the probabilities of the nine outcomes with the probabilities at the first roll in part (a).

   (c) In Yahtzee, often you wish to get 4-of-a-kind and your first roll is only one-pair. To learn about the relationship between the first and final rolls, type the command

```
table 'ROLL_1'*'F_ROLL'
```

How many times did you roll a one-pair on the first roll? Of these rolls, how many resulted in a final roll of 4-of-a-kind? Estimate the conditional probability of getting 4-of-a-kind given a first roll of one-pair.

(d) Repeat part (c) to estimate the probability of getting a large straight on the final roll given that you got a small straight on the first roll.

3. Try playing Yahtzee a large number of times using a particular strategy different from that of the computer. (One alternative possible strategy is to ignore small and large straights and concentrate on getting 3-of-a-kind and 4-of-a-kind.) Describe in words what strategy you will be using. Record the final outcome after each game. After you have played many games, compare the relative frequencies of the nine outcomes. Compare the results with those using the computer strategy.

4. Run the baseball simulation described in Section 2.4 for eleven teams. Let the strengths for the eleven teams be given by equally spaced values from $-.5$ to $.5$. When the simulation is completed, print the columns 'n_win_1'–'n_win_11' and the columns 'place_1'–'place_11'. Each row of the table corresponds to a season. The first entry in 'n_win_1' gives the number of wins for the weakest team for the first season, the entry in 'n_win_2' corresponds to the number of wins for the next weakest team, and so on. The numbers in the columns 'place_1'–'place_11' give the corresponding place finishes.

Using the tally command on 'n_win_11', summarize the number of wins for the best team. What is a typical number of games won? What is the fewest and most games won by this best team? In the twenty seasons, how many times did this best team win the season? (Look at the column 'place_11'.)

Chapter 3

Introduction to Inference Using Bayes' Rule

3.1   Example: Do you have a rare disease?

Uncertainty is a basic aspect of our world. There are many events in our lives in which we are uncertain about the outcome. For example, we may not be certain about tomorrow's weather, where we will be living ten years from now, or whether we have a rare disease. Probabilities provide a means of measuring uncertainty. A probability of a particular event is a number between 0 and 1 that measures one's subjective opinion about the likelihood of the event. Probabilities are conditional in that one's opinion about an event is dependent on our current state of knowledge. As we gain new information, our probabilities can change. Bayes' rule provides a mechanism for changing our probabilities when we obtain new data.

The basic principles of Bayes' rule can be introduced by the following example. Suppose that you are given a blood test for a rare disease. The proportion of people who currently have this disease is .001. The blood test comes back with two results: positive, which is some indication that you may have the disease, or negative. It is possible that the test will give the wrong result. If you have the disease, it will give a negative reading with probability .05. Likewise, it will give a false positive result with probability .05. Suppose that you have a blood test and the result is positive. Should you be concerned that you have the disease?

In this example, you are uncertain if you have the rare disease. There are two possible alternatives or *models*: you have the disease, or you don't have the disease. Before you have a blood test, you can assign probabilities to "have disease" and "don't have disease" that reflect the plausibility of these two models. You think that your chance of having the disease is similar to the chance of a randomly selected person from the population. Thus you assign the event "have disease" a probability of .001. By a property of probabilities, this implies that the event "don't have disease" has a probability of $1 - .001 = .999$. These assigned probabilities are referred to as *prior probabilities* since they reflect your opinions about these two models prior to the blood test.

The new information that we obtain to learn about the different models is called *data*. In this example, the data is the result of the blood test. The different data results are called *observations* or *outcomes*. Here the two possible observations are a positive result $(+)$ or a negative result $(-)$. We are given the probabilities of the observations for each model. If we "have the disease," the probability of a $+$ observation is .95 and the probability of a $-$ observation is .05. Likewise, if we "don't have the disease," the probabilities of the outcomes $+$ and $-$ are .05 and .95, respectively. These probabilities are called *likelihoods*— they are the probabilities of the different data outcomes conditional on each possible model.

All of the numerical information for this problem is placed in the following table. The column MODEL lists the two possible models, and the column PRIOR contains the corresponding prior probabilities. The third and fourth columns of the table contain the likelihoods. The third column lists the probabilities of the first observation $+$ conditional on each model, and the fourth column lists the probabilities of $-$ for each model.

MODEL	PRIOR	LIKELIHOODS	
		$P(+$ given MODEL$)$	$P(-$ given MODEL$)$
Have disease	.001	.95	.05
Don't have disease	.999	.05	.95

Now you take the blood test and the result is positive. Bayes' rule is the formula for changing your probabilities about the models in the light of this data. This rule states that the probability of a model conditional on an observation (the *posterior probability*) is proportional to the product of the prior probability of the model and the likelihood of the observation OBS. That is,

$P(\text{MODEL given OBS})$ is proportional to $P(\text{MODEL}) \times P(\text{OBS given MODEL})$.

Bayes' rule can be implemented for our example in a table format. We are interested in computing the posterior probabilities of the two models "have disease" and "don't have disease" conditional on the positive test result +. The table below summarizes the calculations of the posterior probabilities. The PRIOR and LIKE columns contain the prior probabilities and likelihood values for the + observation. (These are taken from the earlier table.) The PRODUCT column contains the products of the prior probabilities and the likelihoods. The value at the bottom of the PRODUCT column is the sum of the products, and the posterior probabilities in the column POST are obtained by dividing each product by the sum. The number at the bottom of the POST column table is the sum of the posterior probabilities. This value of 1 confirms that this last column contains a valid set of probabilities.

MODEL	PRIOR	LIKE	PRODUCT	POST
Have disease	.001	.95	.00095	.019
Don't have disease	.999	.05	.04995	.981
			.05090	1

What have you learned from the blood test? Before the test, the probability you had the disease was .001. After this positive test result, it has increased to .019. The probability of having the disease has risen dramatically (by a factor of 19), but the actual size of the probability is still small. You would need more evidence to convince yourself that you actually had the disease.

3.2 Using Minitab

> Minitab commands to perform Bayes' rule:
> To set up the problem:
> **exec 'bayes_se'**
> To update probabilities:
> **exec 'bayes'**

To define the model, prior probabilities, and likelihoods for the disease testing example, run the program 'bayes_se'. In the output here, first input the number of models, the names of each model, and the initial probabilities assigned to the models. Next, input the name of each possible observation — here the characters + and − are used to denote the positive and negative test results. Last, enter the likelihoods for each model. Model 1 is the "have disease" model. For this model, enter the probabilities of a + and − test result. Similarly, the probabilities of the two observations are entered for the second "don't have disease" model. The program confirms you have correctly entered the basic information by printing the table of probabilities and likelihoods.

```
MTB > exec 'bayes_se'

INPUT NUMBER OF MODELS:
DATA> 2

INPUT NAMES OF MODELS (ONE NAME ON EACH LINE):
DATA> have disease
DATA> don't have disease

INPUT PRIOR PROBABILITIES OF MODELS:
DATA> .001 .999

INPUT THE NUMBER OF POSSIBLE OUTCOMES:
DATA> 2

INPUT THE NAME OF EACH OBSERVATION:
 (ONE OBSERVATION ON A LINE)
DATA> +
DATA> -

INPUT LIKELIHOODS OF EACH MODEL:
MODEL
   1
DATA> .95 .05

MODEL
   2
DATA> .05 .95
```

TABLE OF PROBABILITIES OF MODELS AND OUTCOMES:

```
ROW  MODEL          NAME  PRIOR  OUT_1  OUT_2
  1      1  have disease  0.001   0.95   0.05
  2      2  don't have d  0.999   0.05   0.95
```

To update the probabilities based on the positive test result, the program 'bayes' is executed. This program asks for the number of observations and the names of the observation. Here one observation was made that is named +. The program then updates the probabilities in the table format described in the previous section.

```
MTB > exec 'bayes'

INPUT NUMBER OF OBSERVATIONS:
DATA> 1

INPUT OBSERVATIONS:
 (ONE OBSERVATION NAME ON A LINE:)
DATA> +

OUTCOME
   +
ROW  MODEL          NAME  PRIOR  LIKE  PRODUCT      POST
  1      1  have disease  0.001  0.95  0.00095  0.018664
  2      2  don't have d  0.999  0.05  0.04995  0.981336
```

3.3 Example: Learning about a student's ability

As a second illustration of the use of Bayes' rule, suppose that a new student is attending a high school, and the principal is uncertain about the student's ability. This administrator classifies students as either "good," "mediocre," or "poor." (These three types of students will represent the *models* in this problem.) The principal knows very little about this particular student, but she is familiar with other students who have transferred in from the same community. Of these transfer students, she believes that 60% were good, 30% were mediocre, and only 10% were poor. The administrator thinks that the new student is representative of other students that have come from this community, and so she believes that the student is good, mediocre, or poor with respective probabilities .6, .3, and .1.

The principal is also knowledgeable about the types of grades in core subjects earned by students of the three types. In these courses, the possible grades are A, B, C, D, and F. For each course, suppose that a "good student" gets either a A, B or C with probabilities .4, .4, .2, respectively. A "mediocre student" gets A, B, C, D, or F, with respective probabilities .1, .2, .4, .2, .1, and a "poor student" will only get a C, D, or F, with probabilities .3, .5, .2. Using the terminology of the previous section, the *data* refers to a particular grade in a course. There are five possible *observations*, and the *likelihoods* are the given probabilities of the five different grades for the three different types of student.

The principal will learn more about the ability of the student after he has taken some classes. Suppose that he takes four classes and the grades in the four courses are independent (which means that the student's chances of a particular grade in one course is not dependent on his performance in other courses). At the end of the term, the principal observes that the student gets two B's and two C's. What does the principal now think about the student's ability?

The macro 'bayes_se' is used to set up this problem. In this example, one has three models named "good student," "mediocre student," and "poor student" that correspond to the three ability levels. The prior probabilities assigned to the three models are the probabilities of the principal before the student took any course work. In this example, an observation corresponds to a particular grade in a course. There are five possible outcomes corresponding to the five grades; the obvious names A, B, C, D, F are used for these observations. Last, one inputs the likelihoods that are the probabilities of the five grades for each model. In this example, one inputs the probabilities of the entire set of outcomes (A, B, C, D, F) when the student is "good," inputs a second set of probabilities when the student is "mediocre," and a third set when the student is "poor." Since there are five outcomes, one must input five numbers per line. If the outcome (grade) is not possible, then a zero probability is entered.

```
MTB > exec 'bayes_se'

INPUT NUMBER OF MODELS:
DATA> 3

INPUT NAMES OF MODELS (ONE NAME ON EACH LINE):
DATA> good student
DATA> mediocre student
DATA> poor student

INPUT PRIOR PROBABILITIES OF MODELS:
DATA> .6 .3 .1

INPUT THE NUMBER OF POSSIBLE OUTCOMES:
DATA> 5

INPUT THE NAME OF EACH OBSERVATION:
 (ONE OBSERVATION ON A LINE)
DATA> A
DATA> B
DATA> C
DATA> D
DATA> F

INPUT LIKELIHOODS OF EACH MODEL:
MODEL
   1
DATA> .4 .4 .2 0 0

MODEL
   2
DATA> .1 .2 .4 .2 .1

MODEL
   3
DATA> 0 0 .3 .5 .2

TABLE OF PROBABILITIES OF MODELS AND OUTCOMES:
```

ROW	MODEL	NAME	PRIOR	OUT_1	OUT_2	OUT_3	OUT_4	OUT_5
1	1	good student	0.6	0.4	0.4	0.2	0.0	0.0
2	2	mediocre stu	0.3	0.1	0.2	0.4	0.2	0.1
3	3	poor student	0.1	0.0	0.0	0.3	0.5	0.2

To compute posterior probabilities after observing two B's and two C's, one executes the macro 'bayes'. The program asks for the number of observations and then to enter each observation on a separate line. In this example, we can view the observed grades as the list B, B, C, C, the list B, C, B, C, or as any arrangement of two B's and two C's. There are four observations, and one enters the observations using the same names as defined in 'bayes_se'.

The program will update the probabilities of the models one observation at a time. The first data outcome is the grade B. The output here shows the computations of the posterior probabilities. Initially, one believed that the student was good, mediocre, or poor with respective probabilities .6, .3, .1. After observing a B, the principal knows that the model "poor student" is no longer possible, and the remaining two models "good" and "mediocre" have probabilities .8 and .2.

```
MTB > exec 'bayes'

INPUT NUMBER OF OBSERVATIONS:
DATA> 4

INPUT OBSERVATIONS:
 (ONE OBSERVATION NAME ON A LINE:)
DATA> B
DATA> B
DATA> C
DATA> C

OUTCOME
  B
```

ROW	MODEL	NAME	PRIOR	LIKE	PRODUCT	POST
1	1	good student	0.6	0.4	0.24	0.8
2	2	mediocre stu	0.3	0.2	0.06	0.2
3	3	poor student	0.1	0.0	0.00	0.0

The posterior probabilities after observing the first observation now become the prior probabilities before observing the next grade. The first table in the output here displays the updating for the second grade B. The prior probabilities that are shown in this table are taken from the POST column in the first Bayes' rule table. The posterior probabilities in the table (.888889, .111111, 0) reflect the principal's beliefs after observing two B's.

OUTCOME
  B

ROW	MODEL	NAME	PRIOR	LIKE	PRODUCT	POST
1	1	good student	0.8	0.4	0.32	0.888889
2	2	mediocre stu	0.2	0.2	0.04	0.111111
3	3	poor student	0.0	0.0	0.00	0.000000

OUTCOME
  C

ROW	MODEL	NAME	PRIOR	LIKE	PRODUCT	POST
1	1	good student	0.888889	0.2	0.177778	0.8
2	2	mediocre stu	0.111111	0.4	0.044444	0.2
3	3	poor student	0.000000	0.3	0.000000	0.0

OUTCOME
  C

ROW	MODEL	NAME	PRIOR	LIKE	PRODUCT	POST
1	1	good student	0.8	0.2	0.16	0.666667
2	2	mediocre stu	0.2	0.4	0.08	0.333333
3	3	poor student	0.0	0.3	0.00	0.000000

SUMMARY OF PRIOR AND POSTERIOR MODEL PROBABILITIES:

ROW	OBS_NO	OUTCOMES	PROB_M1	PROB_M2	PROB_M3
1	0		0.600000	0.300000	0.1
2	1	B	0.800000	0.200000	0.0
3	2	B	0.888889	0.111111	0.0
4	3	C	0.800000	0.200000	0.0
5	4	C	0.666667	0.333333	0.0

In a similar fashion, the probabilities are updated for the final two grades C and C, one grade at a time. The final Bayes' rule table shows that, after observing the four grades, the principal believes that the student is twice as likely to be good than mediocre.

The program concludes by listing the changing posterior model probabilities. The observations are listed in the column OUTCOMES. For each outcome value, the posterior model probabilities are listed in the columns PROB_M1, PROB_M2, etc. (Recall the first model is "good student", the second model is "mediocre" student, and so on.) This table is useful in seeing how the model probabilities change as you obtain more data.

3.4   Exercises

1. (Berry[1], Exercise 5.32) A bowl contains four chips; some are red and some are green. You don't know how many there are of each color, but you believe that there are either one, two, or three reds, and you place the same probability (.333) on each possibility. Suppose that you select two chips from the bowl with replacement and find two reds. You wish to calculate your updated probabilities of the three possibilities. First use the program 'bayes_se' to set up the problem. You have three models (1 red, 2 red, 3 red) with prior probabilities .333, .333, .333. There are two possible outcomes (represent red by the letter r and green by g). For the first model (1 red), the probability of these two outcomes are .25, .75; for the second model (2 red), these outcomes have probabilities .5,.5; and for the third model, the outcomes have probabilities .75, .25. You then can update the probabilities by using the program 'bayes'; here the outcomes are two reds, which you enter into the program as r and r on separate lines.

2. Consider the blood testing example discussed in the chapter. Say that you have three independent blood tests and they are all positive. Do you have a high probability of having the disease? Use the Minitab programs for the computations. Here there are two models (1, have disease; 2, don't have disease) and two possible outcomes (+, positive result; −, negative result). Your observations are +, +, +.

3. (Berry[1], example 5.11) Suppose that you have two dice; one is a four-sider and one is a ten-sider. I choose one of the dice at random and roll it twice, getting a 3 and a 4. What is the probability that I'm rolling the four-sided die? (Here you have two models, four-sided and ten-sided, and since you are choosing a die at random, the probability of each model is .5. An outcome corresponds to the roll of a die. You observe the outcomes 3 and 4.)

---

[1]Berry, D. (1996), *Statistics: A Bayesian Perspective*, Belmont, CA.: Duxbury Press.

4. Berry[2], Chapter 10, describes a carnival game in which you throw darts at a board. The board is subdivided into a large number of small squares labeled with one of the numbers 1, 2, 3, 4, 5, 6. The object of the game is to throw a set number of darts and obtain a large score. Consider three types of players with different abilities. The probabilities that these players will obtain the numbers 1,...,6 on a single throw are given in the following table:

	Outcome					
Player	1	2	3	4	5	6
Poor	.5	.3	.1	.1	0	0
Average	.2	.2	.2	.2	.1	.1
Good	.05	.15	.1	.2	.25	.25

Suppose that you watch a player play the game who is either "poor," "average," or "good," but you have little idea about the ability of the player and assign each possibility the same probability. You watch him play and observe the following dart throws:

$$5, 4, 3, 4, 5, 3, 2, 1, 2, 4$$

What are your updated probabilities about the ability of the player?

5. (Schmitt[3], exercise 3.2.6) A geneticist is investigating the linkage between two fruit-fly genes, and plans a test cross. Under hypothesis A, the proportions of progeny to be expected are

Type 1	Type 2	Type 3	Type 4
.5625	.1875	.1875	.0625

while under hypothesis B, they are

Type 1	Type 2	Type 3	Type 4
.25	.25	.25	.25

However, he doesn't know which hypothesis to expect, so he assigns each hypothesis a prior probability of .5. He observes the progeny of seventeen hatched fruit-flies, with the following frequencies

---

[2]Berry, D. (1996), *Statistics: A Bayesian Perspective*, Belmont, CA.: Duxbury Press.

[3]Schmitt, S. A. (1969),*Measuring Uncertainty: An Elementary Introduction to Bayesian Statistics*, New York: Addison-Wesley.

Type 1	Type 2	Type 3	Type 4
8	2	4	3

What are the posterior probabilities of the two hypotheses?

In this problem, the two models correspond to the two hypotheses and the four outcomes are the four types of progeny. The data are the progeny of the seventeen flies, which are

$$1, 1, 1, 1, 1, 1, 1, 1, 2, 2, 3, 3, 3, 3, 4, 4, 4.$$

6. (Schmitt[4], example 3.2.1) An automatic machine in a small factory produces metal parts. Most of the time (90% by long records), it produces 95% good parts and the remaining have to be scrapped. Other times, the machine slips into a less productive mode and only produces 70% good parts. The foreman observes the quality of parts that are produced by the machine and wants to stop and adjust the machine when she believes that the machine is not working well. Suppose that the first dozen parts produced are given by the sequence

$$s, u, s, s, s, s, s, s, s, u, s, u,$$

where s = satisfactory and u = unsatisfactory. After observing this sequence, what is the probability that the machine is in its good state? If the foreman wishes to stop the machine when the probability of "good state" is under .7, when should she stop?

In the use of the programs, the two possible models are "good" and "bad," a part produced has two outcomes "satisfactory" (recorded as an s) and "unsatisfactory" (recorded as a u), and you observe the sequence of letters given. When the program 'bayes' is run, then the probability of the model "good" will be given after each observation. By looking at this list of probabilities, one can tell when the foreman should stop.

---

[4]Schmitt, S. A. (1969), *Measuring Uncertainty: An Elementary Introduction to Bayesian Statistics*, New York: Addison-Wesley.

Chapter 4

Learning about a Proportion

4.1 Introduction

This chapter describes the use of a set of Minitab programs to learn about a population proportion. The population is the collection of people or units that is the object of our inference. Suppose that the population can be divided into two groups that will be referred to as "successes" and "failures." The proportion of successes in the population will be denoted by p. To learn about the value of the proportion, some data are obtained. A random sample of units is taken from the population; suppose that one observes s successes and f failures in this sample.

A basic inference problem is to learn about the unknown value of the proportion p from the sample outcome of s successes and f failures. In Bayesian inference, after seeing the data, all information about the unknown proportion is contained in the probability distribution of p, the *posterior distribution*. This posterior distribution can be summarized to obtain different types of inferences. One may be interested in guessing or estimating the value of p. This *estimate* will be a single value of p that has a large probability under the posterior distribution. A second type of inference is an interval estimate for p. This is a interval of values between 0 and 1 that contains a high probability content of the posterior distribution. In this book, we will refer to this interval estimate as a *probability interval*.

Estimates and probability intervals are generally useful for learning about the location of a proportion. In some problems, one is also interested if certain sets of values for p are plausible. For example, one may wonder whether the proportion p exceeds .8. One decides whether this statement is likely by computing the probability $P(p > .8)$ from the posterior distribution. If this probability is sufficiently small, say, smaller than .01, then one can reject this statement and conclude (with 99% confidence) that $p \leq .8$.

All of the inferences just described focused on the population proportion p. A different type of inference is to learn about future observations. Suppose that a new random sample is drawn from the population of a particular size. What is the probability that one obtains t successes and u failures in this new sample? This is typically called a prediction problem to be distinguished from the estimation problem of p described earlier.

All of the programs of this chapter will be illustrated by the following example. Suppose that a new baseball player is joining the major leagues. The manager of the team is interested in learning about the player's batting average p. To put this example in the previous framework, we suppose that the player has an infinite number of opportunities to bat, and each opportunity in this hypothetical population is classified as a hit (success) or an out (failure). Then the batting average p is the proportion of hits in this infinite population. (Note that this definition is different from the usual interpretation of a batting average in baseball. A batting average is typically viewed as the observed fraction of hits during a season. Here we are viewing a batting average as a measurement of the player's true batting ability.) Suppose that the hitter plays five games, has twenty at-bats, and gets five hits. The manager is interested in learning about the player's batting average on this basis of this hitting performance. Also, since the player's performance in future games is also important, the manager would like to predict the number of hits of the player in the next twenty at-bats.

In developing Bayesian methods to answer these inference problems, a prior distribution needs to be constructed for the unknown proportion p. There are two basic methods for constructing a prior that will be illustrated in the following sections. The *discrete* method builds a prior distribution by first selecting a finite set of values of p and then assigning probabilities to each of the values of p in the set. The *continuous* method assumes that p is continuous valued on the interval from 0 to 1 and constructs a density function on this interval that reflects one's prior information about this proportion.

4.2 Using discrete models

4.2.1 Constructing a prior distribution

In this section, we illustrate learning about a proportion p when there are a finite collection of values of p of interest. The first step is to construct a prior distribution. This distribution consists of a set of proportion values and a set of associated probabilities that reflect the opinion of the user about the likelihood of the different values. To obtain this distribution, one must first select a plausible set of proportion values. In some applications, the value of the proportion may be known to fall into a particular interval, and one can choose five to ten equally spaced values for p within this interval. If one has little knowledge of the location of the proportion, then an equally spaced grid of values from 0 to 1 can suffice.

In our example, the manager is interested in the player's batting average p. In baseball, season batting averages of full-time players fall in the interval $(.2, .35)$. In baseball, particular values of p in this interval range have special meaning. A proportion value $p = .2$ is representative of a relatively weak hitting player, $p = .25$ is mediocre, $p = .3$ reflects a pretty good hitter, and $p = .35$ is exceptional. Since the particular player's hitting ability is not known, all of these four states are possible, and so the manager chooses the set of proportion values $\{.2, .25, .3, .35\}$.

Next, one assigns probabilities to the different values of p that reflect one's knowledge about the proportion. One method of doing this is to think first of the most likely value of p, and assign it a large positive number. Then one thinks about other proportion values in relation to the most likely value. After performing a number of these comparisons, a probability distribution can be found that roughly approximates one's opinion about the proportion.

In our baseball example, the manager can construct a prior distribution based on his knowledge of players in the past who have started playing in the major leagues. He knows that most hitters who start playing are mediocre with an approximate batting average of .25. So he initially assigns the value $p = .25$ the large number 100. Then he thinks of the remaining values $p = .2, .3, .35$. It is relatively common for hitters to be weak when they start in the majors, but he thinks that "mediocre" is twice as likely as "weak." So he assigns $p = .2$ the value of 50. Good hitters are relatively rare and exceptional hitters even more so. He thinks it is five times more likely that the player is "mediocre" than "good"; similarly he thinks it is five times more likely that the player is "good" compared to "exceptional." So the values of $p = .3$ and $.35$ are assigned the numbers 20 and 4, respectively. The sum of the numbers assigned is 174. To make these assigned numbers probabilities, one divides each number by the sum. The resulting probability distribution on $p = .2, .25, .3, .35$ is given by $50/174$, $100/174, 20/174, 4/174$, respectively.

In some situations, it can be difficult to construct a prior distribution on the values of p. In cases where one is reluctant to construct a subjective probability distribution or indifferent between the different proportion values, one can use a noninformative prior distribution. This distribution assigns to each proportion value the same probability. In our baseball example, since there are four values of interest, a noninformative prior would assign a probability $1/4 = .25$ to each of the values $p = .2, .25, .3, .35$.

4.2.2 Computing the posterior distribution

> Minitab command using discrete models for a proportion:
> Learning about p:
> **exec 'p_disc'**

After a prior probability distribution has been constructed, one wishes to update these probabilities after observing data. This updating can be performed as a basic application of Bayes' rule. In this inference setting, the value of the proportion is unknown and one can think of a "model" as a particular value of p. To illustrate, suppose that one decides to use eleven equally spaced values of the proportion from 0 to 1. Then there are eleven possible models, $p = 0, p = .1, \ldots, p = 1$. The observation is (s, f), the number of successes and failures in the random sample. For a value of p, the likelihood is the probability of this data outcome. Think of the data outcome as a particular sequence of successes and failures, such as

$$S, S, \ldots, S, F, F, \ldots, F,$$

where S denotes a success, F denotes a failure, and the number of S's (F's) in the sequence is s (f). Then the likelihood of this observation is given by the product

$$p \times p \times \ldots \times p \times (1 - p) \times (1 - p) \times \ldots \times (1 - p),$$

p	PRIOR	LIKE	PRODUCT	POST
0	$P(0)$	$0^s(1-0)^f$	$P(0)0^s(1-0)^f$	$P(0)0^s(1-0)^f$ / SUM
.1	$P(.1)$	$.1^s(1-.1)^f$	$P(.1).1^s(1-.1)^f$	$P(.1).1^s(1-.1)^f$ / SUM
.2	$P(.2)$	$.2^s(1-.2)^f$	$P(.2).2^s(1-.2)^f$	$P(.2).2^s(1-.2)^f$ / SUM
$\vdots$	$\vdots$	$\vdots$	$\vdots$	$\vdots$
1	$P(1)$	$1^s(1-1)^f$	$P(1)1(1-1)^f$	$P(1)1(1-1)^f$ / SUM
			SUM	1

Table 4.1: Computation of posterior probabilities for proportion problem.

which is equal to $p^s(1-p)^f$. The posterior probability of the value p is proportional to the product of its prior probability and this likelihood value. The computation of these posterior probabilities is illustrated in Table 4.1. Note that for each value of p, the table lists the prior probability $P(p)$, the likelihood, and the product. The set of posterior probabilities is found by dividing each product by the sum of the products.

These computations are illustrated for the baseball example with a noninformative prior using the program 'p_disc'. Recall the basic information for this example — there are four p models, equal probabilities are assigned to the four proportion values, and a data set of five successes and fifteen failures is observed. Before this program is run, one must name two columns 'p' and 'prior'. The set of proportion values is placed in the column 'p', and the corresponding probabilities are placed in the column 'prior'. In the output, these numbers are placed into the worksheet using two Minitab set commands. After these two columns are defined, one can run the program 'p_disc'. The user inputs the numbers of successes and failures. The output of this program is the computation of the posterior probabilities in table form. The column 'prior' contains the prior probabilities, the column 'LIKE' the likelihoods, the column 'PRODUCT' the respective products, and the posterior probabilities are placed in the column 'POST'. (The column 'POST' is saved on the worksheet for future computations.) To summarize the sets of probabilities, two columns 'P_x_PRIO' and 'P_x_POST' are added. The column 'P_x_PRIO' contains the respective products of the proportions and the prior probabilities; the sum of this column, placed at the bottom, is the prior mean of the distribution. Likewise, the sum of the 'P_x_POST' column, placed at the bottom, is the mean of the posterior probabilities. The posterior mean can be interpreted as the probability of a success in a single future observation. This program also graphs the posterior probabilities using a line graph (see Figure 4.1).

```
MTB > name c1 'p' c2 'prior'
MTB > set 'p'
DATA> .2 .25 .3 .35
DATA> end
MTB > set 'prior'
DATA> .25 .25 .25 .25
DATA> end
MTB > exec 'p_disc'

INPUT OBSERVED NUMBER OF SUCCESSES AND FAILURES:
DATA> 5 15

  ROW      p   prior  P_x_PRIO      LIKE  PRODUCT       POST  P_x_POST
    1   0.20    0.25    0.0500    862742   215686   0.255595  0.051119
    2   0.25    0.25    0.0625   1000000   250000   0.296259  0.074065
    3   0.30    0.25    0.0750    884011   221003   0.261897  0.078569
    4   0.35    0.25    0.0875    628668   157167   0.186249  0.065187
    5                   0.2750                                0.268940
```

How have these data changed the manager's opinion of the batting ability of the player? It has slightly dropped. The most likely value of p, the value with the largest probability, is .25. The probability that the player is an exceptional hitter is .186, down from the prior probability of .25. The posterior mean, .268, is slightly smaller than the prior mean of .275. The probability that the hitter will get a hit in the next at-bat is .268.

4.2.3   Summarizing a posterior probability distribution

> Minitab command for summarizing a discrete probability distribution:
> **exec 'disc_sum'**

The program 'disc_sum' is useful for graphing and summarizing posterior probabilities for a proportion. To illustrate different types of summaries, consider a baseball example related to the one discussed in this chapter. In 1994, Paul O'Neill of the New York Yankees had an excellent batting season, obtaining 132 hits in 368 at-bats for a batting average of 132/368=.359. Suppose that one regards the 368 at-bats as independent

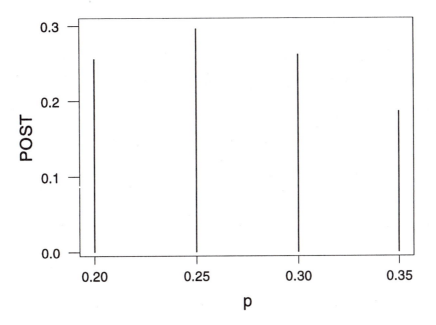

Figure 4.1: Posterior probabilities for proportion p in baseball example.

trials with a constant probability p of getting a hit on a single trial. We can view p as O'Neill's true batting average; this is the proportion of hits if he were able to take an infinite number of at-bats in 1994.

What have we learned about O'Neill's true batting average on the basis of his hitting performance? We use the program 'p_disc' to find the posterior probabilities for p. We place in the column 'P' the fine grid of values .2, .21, ..., .5 and let the prior distribution be uniform over this grid. The program 'p_disc' is run with the observed data: $s = 132$ hits and $f = 236$ outs. The program saves the posterior probabilities in the column 'POST'.

In the output, the macro 'disc_sum' is executed to summarize this discrete posterior distribution. There are two inputs to this program: the numbers of the columns that contain the values of the variable and the probabilities. The values of 'P' were stored in column C1 of the worksheet, and the program 'p_disc' places the posterior probabilities in column C51. (You can check this by use of the Minitab info command.) So the values 1 and 51 are inputted in the program 'disc_sum'.

```
MTB > exec 'disc_sum'

INPUT NUMBER OF COLUMN WHICH CONTAINS VALUES OF VARIABLE:
DATA> 1

INPUT NUMBER OF COLUMN WHICH CONTAINS PROBABILITIES:
DATA> 51

TYPE 'y' TO SEE A PLOT OF THE PROBABILITIES:
y

TYPE 'y' TO GET SUMMARIES OF THE DISTRIBUTION:
y

 Row    MODE        MEAN         STD
   1    0.36     0.359460    0.0249122

TYPE 'y' TO COMPUTE CUMULATIVE PROBABILITIES:
--------------------------------------------------------------------
   Input values of variable of interest.  The output is the column of
   values and the column of cumulative probabilities PROB_LT.
--------------------------------------------------------------------
y
DATA> .2 .39
DATA> end

 Row    VALUE     PROB_LE
   1    0.20     0.000000
   2    0.39     0.922897

TYPE 'y' TO COMPUTE PROBABILITY INTERVALS:
--------------------------------------------------------------------
   Input list of probabilities.  For each probability p, the
   set of values of the variable for which the probability content
   of the set exceeds p is given.
--------------------------------------------------------------------
y
DATA> .5 .9
DATA> end
```

```
PROB_SET
  0.574893

SET
  0.34   0.35   0.36   0.37
--------------------------------------------------------------------


PROB_SET
  0.931126

SET
  0.32   0.33   0.34   0.35   0.36   0.37   0.38   0.39   0.40
--------------------------------------------------------------------
```

One can first plot the probability distribution. The plot of the probabilities (not shown) resembles the one displayed by the macro 'p_disc'. Next, the program gives some summary statistics for the distribution. The *mode* is the value of the proportion p with the largest probability. The *mean* is a weighted average of the values of p. Here the mode is .36 and the mean is .359. These two measures of center are approximately equal, which indicates that the probability distribution is symmetric. The program also displays the *standard deviation* .0249 of the distribution. This is a measure of the spread of the probability distribution. If a fine grid of values is used and the distribution is roughly bell-shaped, then approximately 95% of the total probability is located within two standard deviations of the mean.

The program will also compute *cumulative probabilities* of the distribution. In our example, suppose that we're interested in the probability that O'Neill's 1994 true batting average is no larger than .200. Equivalently, we're interested in the cumulative probability $P(p \leq .2)$. In addition, we're interested in the chance that his true average is .400 or higher. This probability is expressible as a cumulative probability since $P(p \geq .4) = 1 - P(p < .4) = 1 - P(p \leq .39)$. (The largest possible value of the proportion smaller than .4 is .39.) In the printout, we indicate by typing y that we wish to compute cumulative probabilities and enter the values .2 and .39 on the DATA line. For each value, the program gives the cumulative probability. We see that $P(p \leq .2)$ is approximately 0, indicating that O'Neill's true batting average for 1994 can't be .200 or smaller. However, $P(p \geq .4) = 1 - .923 = .077$ — there is a small chance that O'Neill was truly a .400 hitter that year.

The macro 'disc_sum' can also be used to compute *probability intervals.* For a given probability value *PROB*, the program will find the set of mostly likely values of the proportion p such that the probability content of the set will be equal to or greater than *PROB*. Here the values .5 and .9 are inputted on the DATA line. We see from the output that the set {.34, .35, .36, .37} has a probability content of .575 and the probability of the set {.32, .33, ..., .40} is .931. Thus we are 93% confident that O'Neill's batting average in 1994 was between .32 and .40. It may be surprising to note that one is still fairly uncertain about a proportion value even after observing a sample of size 368.

4.2.4   Predicting the number of successes in a future experiment

> Minitab command using discrete models for a proportion:
> Learning about a future experiment:
> **exec 'p_disc_p'**

The program 'p_disc' focuses on learning about the proportion p. A second type of inference is about the number of successes and failures in a future experiment. In the baseball example, the manager may be interested about the player's performance in twenty future at-bats. How many hits will he get? What's the most likely number of hits?

The computation of the *predictive* probability of t successes and u failures in a future experiment is illustrated in Table 4.2. As before, there is a set of proportion values of interest. Here we use equally spaced values of p between 0 and 1. The next column, PROB, contains the current probabilities for these proportion values. These probabilities could be either *prior* probabilities before any data is observed or *posterior* probabilities after taking some observations. The column LIKE contains the probability of observing t successes in a future experiment of $t+u$ trials if the probability of a *success* is given by p. This number is the binomial probability

$$\text{LIKE} = \frac{(t+u)!}{t!u!}p^t(1-p)^u.$$

In the table, the symbol C is equal to the binomial coefficient $\frac{(t+u)!}{t!u!}$. The column PRODUCT contains the products of the probabilities and the binomial likelihoods. The sum of the numbers, PRED, is the predictive probability of interest. Typically,

p	PROB	LIKE	PRODUCT
0	P(0)	$C\ 0^t(1-0)^u$	$C\ P(0)0^t(1-0)^u$
.1	P(.1)	$C\ 1^t(1-.1)^u$	$C\ P(.1).1^t(1-.1)^u$
.2	P(.2)	$C\ .2^t(1-.2)^u$	$C\ P(.2).2^t(1-.2)^u$
$\vdots$	$\vdots$	$\vdots$	$\vdots$
1	P(1)	$C\ 1^t(1-1)^u$	$C\ P(1)1^t(1-1)^u$
			PRED

Table 4.2: Computation of predictive probabilities for proportion problem.

p	PROB	LIKE	PRODUCT
.2	.25	.1746	(.25) (.1746)
.25	.25	.2023	(.25) (.2023)
.3	.25	.1789	(.25) (.1789)
.35	.25	.1272	(.25) (.1272)
			.1707

Table 4.3: Computation of predictive probability of 5 successes for baseball example.

one is interested in computing this predictive probability for all possible numbers of successes. If, for example, there are twenty future trials, then the number of successes t could be any integer value from 0 to 20.

To illustrate these calculations for an example, suppose that the manager would like to compute the probability that the player will get exactly five hits in the next twenty at-bats. As before, the possible proportion values are .2, .25, .3, .35. The manager hasn't seen the player hit yet, so his current probabilities are the respective prior probabilities .25, .25, .25, .25. Table 4.3 illustrates the computation of the predictive probability of five hits in twenty at-bats. The LIKE column contains the probabilities of five hits in twenty at-bats for each of the proportion values. The sum of the column PRODUCT, .1707, is the predictive probability.

If one was interested in a different predictive probability, say ten hits in the next twenty at-bats, then one would repeat this calculation using different binomial probabilities in the LIKE column. Also, if the manager has already seen the player get five hits in twenty attempts in previous games, then the PROB column would be replaced by the manager's current posterior probabilities.

Predictive probabilities when one has a discrete set of proportion models can be computed using the program 'p_disc_p'. The program assumes that the proportion values are contained in a Minitab column 'P' and the current probabilities for the proportion are contained in either the column 'PRIOR' or the column 'POST', depending if one wishes to compute predictive probabilities based on the prior or the posterior distributions. In the run of the program described next, the program asks the user to specify if 'PRIOR' or 'POST' probabilities are to be used. Then it asks for the number of trials in the future sample and the range of values (low and high) for the number of successes for which predictive probabilities will be computed. In this baseball example, suppose that one wishes to use the prior probabilities, there are twenty future at-bats, and one is interested in numbers of successes in the complete range 0 to 20.

```
MTB > exec 'p_disc_p'

INPUT 1 IF PROBABILITIES ARE IN 'PRIOR' OR
      2 IF PROBABILITIES ARE IN 'POST':
DATA> 1

INPUT NUMBER OF TRIALS:
DATA> 20

INPUT RANGE (LOW AND HIGH VALUES) FOR NUMBER OF SUCCESSES:
DATA> 0 20

PREDICTIVE DISTRIBUTION OF NUMBER OF SUCCESSES:

ROW   SUCC      PRED
  1      0   0.003920
  2      1   0.021895
  3      2   0.060422
  4      3   0.110780
  5      4   0.153032
  6      5   0.170739
  7      6   0.160145
  8      7   0.128905
  9      8   0.089699
 10      9   0.053915
 11     10   0.027846
```

12	11	0.012266
13	12	0.004565
14	13	0.001420
15	14	0.000364
16	15	0.000075
17	16	0.000012
18	17	0.000002
19	18	0.000000
20	19	0.000000
21	20	0.000000

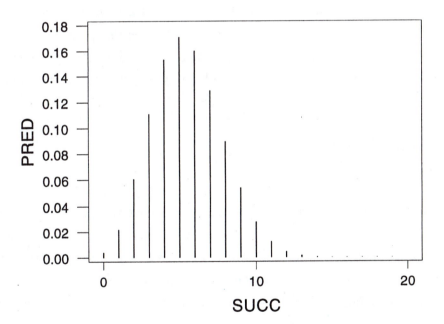

Figure 4.2: Plot of predictive probabilities using a discrete prior for baseball example.

The output of this program is the predictive distribution of the number of hits for the range of values desired. The values of the number of successes and the corresponding predictive probabilities are stored in the Minitab columns 'SUCC' and 'PRED', respectively. The program graphs the predictive probabilities using a line plot (see Figure 4.2).

From this output, what has the manager learned about the number of hits of the baseball player in the next twenty at-bats? The most likely number of hits is five with a predictive probability of .1707. One also notes that numbers of hits close to five also receive relatively large predictive probabilities. In this case, it may be more useful to construct an interval of values that contain a large portion of the entire predictive distribution. The program 'disc_sum' described in Section 4.2.2 can be used to find this probability interval. The column 'SUCC' is C52, and the column 'PRED' is C53; the numbers of these columns are inputs in this program. Suppose that one is interested in finding an interval of values with probability content at least .9. One inputs the value .9 in the program 'disc_sum' and learns from the output of this program that the probability that the number of hits is in the set $\{2, 3, 4, 5, 6, 7, 8, 9\}$ is approximately .93. The manager can be pretty confident (with probability .93) that the player will get between two and nine hits in the next twenty at-bats.

4.3   Using continuous models

4.3.1   Introduction

As in the first part of this chapter, we are interested in learning about a proportion p. What is different here is how we represent our prior information about p. Instead of thinking about a set of possible values for the proportion, we view p as being continuous-valued from 0 to 1. To represent prior opinion about p in this continuous case, we cannot assign probabilities to the individual values of the proportion. However, we may believe that certain regions of the interval $(0, 1)$ are more or less likely than other regions, and we can construct a probability density $g(p)$ to reflect this knowledge. This density is called a prior density, since it represents our opinion about the location of p before any data are collected. Next, some data are collected, and we observe s successes and f failures. Based on these data, our beliefs about p are modified. The adjusted opinions about p are reflected in the posterior density $h(p$ given $(s, f))$. By Bayes' rule, this density is obtained (up to an proportionality constant C) by multiplying the prior density $g(p)$ by the probability of the sample result (s, f) given a value of p:

$$h(p \text{ given } (s, f)) = Cg(p)P((s, f) \text{ given } p) = Cg(p)p^s(1 - p)^f.$$

When one is interested in learning about a continuous-valued proportion p, the usual functional form for a prior density is the beta, with density proportional to $p^{a-1}(1 - p)^{b-1}$. The user chooses values of the beta parameters a and b that approximately reflect his/her information about the unknown proportion. The choice of a beta prior density is convenient in that the posterior density has also a beta functional form. If a beta(a, b) is the prior density for p, and if a binomial experiment results in s successes and f failures, then the posterior density for p is also of the beta form with new parameters $a_1 = a + s$ and $b_1 = b + f$.

The Minitab programs 'beta_sel', 'p_beta' and 'p_beta_p' discussed in this chapter are useful for assessing a beta prior density that approximates one's subjective opinion about p. The program 'beta_sel' will find the particular beta density that matches two statements about the chances of successes in a future sample. The programs 'p_beta' and 'p_beta_p' will perform computations for beta and beta predictive distributions, respectively. These two distributions are useful in checking whether the beta prior density found in 'beta_sel' is a reasonable match to one's prior opinion.

If a beta density is used and data are obtained, one learns about the proportion p by summarizing the beta posterior density. The program 'p_beta' can be used to summarize this beta posterior density. In addition, the program 'p_beta_p' can be used to compute probabilities for numbers of successes in a future data set.

We will describe the assessment of a beta prior distribution and different types of posterior inferences in the same setting that was used for learning about a proportion using discrete models. A baseball manager is interested in learning about the batting average p of a second new major league player. From the manager's knowledge of baseball, he knows that batting averages for most players fall between .2 and .35, and all continuous values inside that interval are possible. The first step in this learning process is to choose a beta density that matches one's prior knowledge about the batting ability of the player. Next, the manager will observe the batting performance of the game for a few games — suppose the hitter gets four hits in forty at-bats.

To update the manager's prior opinion, the posterior density for the probability p will be computed. From this posterior distribution, a number of inferences are possible, three of which will be illustrated by the Minitab programs. One may be interested in a "best" guess at the batting average of the player. This value will be found by finding an average quantity of the posterior density. In addition, the manager would like to find an interval in which he is pretty confident the batting average lies. A *probability* interval will be an interval of values that contain a high probability content of the posterior density. A third type of inference is predicting the player's batting performance in a few games in the future. Suppose the batter gets to bat twenty times during the next three games. What is the most likely number of hits?

Although the proportion values are continuous, particular values of p may be of special interest. In our baseball example, if the player is believed to be a very strong hitter, then one may be interested in the probability that he is a .400 hitter, since .400 is a standard of excellence for a hitter. The probability of this particular proportion value is zero if a continuous density distribution for p is used. However, one can construct a special probability distribution that places a nonzero probability at the particular proportion value of interest. The Minitab program 'p_beta_t' constructs such a special prior distribution, and this distribution can be used to construct a Bayesian test of the statement that $p = .4$.

Although the beta density is a convenient choice for a prior density, one should question whether this particular choice of functional form is a good match to one's opinion about the continuous-valued proportion. In some situations, one may be able only to specify probabilities of intervals and it is not clear that a beta density is a good match to this prior information. In this case, the Minitab program 'p_hist_p' provides a posterior analysis using a more general type of prior information. Suppose that the interval (0, 1) can be partitioned into a small number of intervals and one can assign probabilities to each of the intervals. Then the program will compute the posterior probabilities of the same intervals. The posterior analysis using this *histogram prior* and a beta posterior density will be similar if the two prior densities are similar in shape or if the information in the data is much greater than the information in the prior.

4.3.2 Assessing a beta distribution

> Minitab command for a proportion with continuous models:
> Choosing a beta density based on predictive statements:
> **exec 'beta_sel'**

The Minitab macro 'beta_sel' finds a beta distribution for the baseball manager that matches two prior assessments (see the output below). Instead of directly inputting information about the prior distribution of p, the program asks the manager about the chance of the player hitting in future at-bats. These statements about the predictive distribution of the hit distribution will indirectly give values for the beta parameters a and b. First, the program asks for the probability of a success (hit) on the first at-bat. This number is essentially a guess at the proportion value. Second, if the player does get a hit on the first at-bat, the program asks for the probability of a hit on the second at-bat. This hit on the single at-bat should make one more confident about the hitter's batting ability. So the assessed probability of a second hit given a first hit should be larger than the initial probability of a first hit. In this example, the manager believes that the player will have a successful at-bat with probability .3, and, if he gets a hit, the probability of a second hit is .32. The macro outputs the parameters a and b of the matching beta distribution. In this example, the two assessed probabilities are matched with a beta distribution with parameters 10.2 and 23.8.

```
MTB > exec 'beta_sel'

What is the probability of a success on the first trial?
DATA> .3

If the first trial is a success, what is the conditional probability
of a success on the second trial?
(This should be larger than the first number you gave.)
DATA> .32

The matching values of the beta parameters a and b corresponding
to your predictive probabilities are given by:

 ROW       a      b
   1   10.200   23.8
```

4.3.3   Checking the assessment

> Minitab commands for a proportion with continuous models:
> Summarizing a beta density:
> **exec 'p_beta'**
> Summarizing a beta predictive density:
> **exec 'p_beta_p'**

Although a person can make these two assessments, it is not clear that the matching beta density provides a good approximation to the person's opinion about the proportion p. It is important to check to see whether this particular beta distribution is a reasonable approximation to one's prior beliefs. One can summarize the assessed prior density by the program 'p_beta'; the output for a run of this program is shown later. The inputs for the program are the two beta parameters a and b. The program first graphs the beta density that was assessed. This plot graphically shows what batting averages p are plausible for this prior distribution. This program will also compute cumulative probabilities and percentiles for the beta distribution. To compute cumulative probabilities, the user first enters y when the question about computing these quantities is asked. Then one places a list of proportion values on the DATA line. (An "end" command is entered on a separate line to indicate that no more values are to be entered.) For each proportion value in the list, the probability that p is less than or equal to the value is printed. One can compute percentiles of the beta density in a similar fashion. One inputs y to indicate that one wishes to compute percentiles and enters probability values on the DATA line. For each probability value, the program prints values of the corresponding beta percentiles. The .5 percentile is the median of the probability distribution, and the .25 and .75 percentiles are the lower and upper quartiles of the distribution, respectively.

For the baseball example, the manager wonders whether a beta(10.2, 23.8) prior is a reasonable match to his opinion about the hitting ability of the player. So he uses the program 'p_beta' to plot the beta density to find the probabilities that p is smaller than .1, .2, .3, .4, .5, and to find the .05, .25, .5, .75, .95 percentiles of the distribution. Figure 4.3 shows a graph of the beta density. From the text output, we see that the probability that $p < .2$ is .09 and the probability that $p < .4$ is .89. In addition, the median of the distribution of p is .296, p falls in the interval (.245, .351) with probability .5, and p lies in (.179, .434) with probability .9.

```
MTB > exec 'p_beta'

INPUT VALUES OF BETA PARAMETERS A AND B:
DATA> 10.2 23.8

TYPE 'y' TO SEE A PLOT OF THE BETA DENSITY:
y

TYPE 'y' TO COMPUTE CUMULATIVE PROBABILITIES:
----------------------------------------------------------------
  Input values of P of interest.  The output is the column of
  values P and the column of cumulative probabilities PROB_LT.
----------------------------------------------------------------
y
DATA> .1 .2 .3 .4 .5
DATA> end

  ROW     P      PROB_LT
    1     0.1    0.000793
    2     0.2    0.092903
    3     0.3    0.519992
    4     0.4    0.894790
    5     0.5    0.991727

TYPE 'y' TO COMPUTE PERCENTILES:
----------------------------------------------------------------
  Input probabilities for which you wish to compute percentiles.
  The output is the probabilities in the column PROB and the
  corresponding percentiles in the column PERCNTLE.
----------------------------------------------------------------
y
DATA> .05 .25 .5 .75 .95
DATA> end

  ROW    PROB    PERCNTLE
    1    0.05    0.179486
    2    0.25    0.244887
    3    0.50    0.296039
    4    0.75    0.350879
    5    0.95    0.434082
```

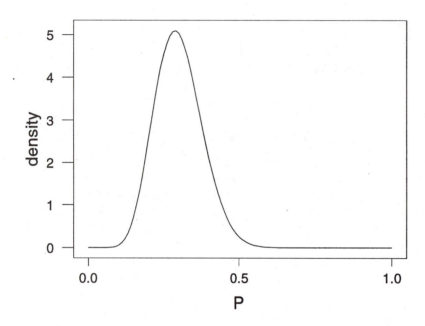

Figure 4.3: The beta (10.2, 23.8) prior density for baseball example.

After computing these summary values of this beta distribution, the user may think harder about his choice of prior. In response to this thinking, the user can make adjustments to the values of the beta parameters a and b that were chosen. In this example, suppose that the manager is fairly certain that the player is not a .400 hitter. Then, the 90% probability interval (.179, .434) will seem too wide for p, and the manager will want to choose an alternative prior distribution that has a smaller spread. A new beta distribution with a smaller spread and approximately the same location can be chosen by multiplying the parameters a and b by the same positive number larger than one. He can now rerun the program 'p_beta' using values of a and b that are double the original values ($a = 20.4$, $b = 47.6$). The output of this program (not shown) tells us that the new 90% probability interval is (.213, .394). The chance that the hitter is a .400 or greater hitter is very small, which may better approximate the user's prior beliefs about the hitting probability p.

The program 'p_beta' is useful in summarizing the beta distribution for the proportion p. One can also check the suitability of the beta distribution by inspection of the predictive distribution of a future binomial experiment. The macro 'p_beta_p' can be used for this purpose. The output of this program is shown next. One inputs the values of the beta parameters a and b and the number of trials n in a future experiment. One also inputs the range of the number of successes s for which predictive probabilities are desired. The macro outputs a table of the number of successes and the respective predictive probabilities. These two columns are stored in the Minitab columns 'SUCC' and 'PRED'. In addition, the predictive probabilities are graphed using a line plot. For the baseball example, this program is run using the beta parameters 20.4 and 47.6. The graph of the predictive probabilities is shown in Figure 4.4. If the batter has twenty opportunities to bat in the future, the probability distribution for the number of hits is given. Note that, for this prior, the most likely numbers of hits are five and six.

```
MTB > exec 'p_beta_p'

INPUT VALUES OF BETA PARAMETERS A AND B:
DATA> 20.4 47.6

INPUT NUMBER OF TRIALS:
DATA> 20

INPUT RANGE (LOW AND HIGH VALUES) FOR NUMBER OF SUCCESSES:
DATA> 0 20

PREDICTIVE DISTRIBUTION OF NUMBER OF SUCCESSES:

    ROW   SUCC       PRED
      1      0   0.002119
      2      1   0.012982
      3      2   0.040232
      4      3   0.083703
      5      4   0.130884
      6      5   0.163249
      7      6   0.168285
      8      7   0.146624
      9      8   0.109537
     10      9   0.070782
```

11	10	0.039741
12	11	0.019405
13	12	0.008219
14	13	0.003001
15	14	0.000935
16	15	0.000245
17	16	0.000052
18	17	0.000009
19	18	0.000001
20	19	0.000000
21	20	0.000000

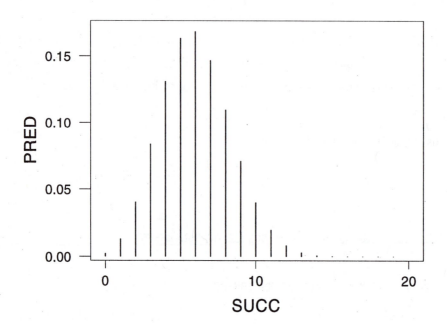

Figure 4.4: Predictive probabilities for a beta(20.4, 47.6) prior density for baseball example.

4.3.4   Learning about the proportion and a future binomial experiment

> Minitab commands for a proportion with continuous models:
> Summarizing a beta density:
> **exec 'p_beta'**
> Summarizing a beta predictive density:
> **exec 'p_beta_p'**

If a beta (a, b) prior distribution is chosen for p and s successes and f failures are observed, then the posterior distribution for p is also beta with parameters $a + s$ and $b + f$. The programs 'p_beta' and 'p_beta_p' discussed in the previous section are also useful in drawing inferences about the beta posterior and beta predictive distributions. To illustrate, consider the earlier example in which a beta (20.4, 47.6) distribution was chosen for the hitting probability p. Suppose that the hitter gets forty chances to bat and only gets four hits. Since $s = 4$ and $f = 36$, the updated distribution for p is beta (24.4, 83.6). The program 'p_beta' is run for this particular example in the output here. The graph of the beta posterior density is given in Figure 4.5.

```
MTB > exec 'p_beta'

INPUT VALUES OF BETA PARAMETERS A AND B:
DATA> 24.4 83.6

TYPE 'y' TO SEE A PLOT OF THE BETA DENSITY:
y

TYPE 'y' TO COMPUTE CUMULATIVE PROBABILITIES:
----------------------------------------------------------------
  Input values of P of interest.  The output is the column of
  values P and the column of cumulative probabilities PROB_LT.
----------------------------------------------------------------
y
DATA> .1 .2 .3 .4 .5
DATA> end

  ROW     P    PROB_LT
    1    0.1   0.00008
    2    0.2   0.26784
    3    0.3   0.96109
    4    0.4   0.99994
    5    0.5   1.00000
```

```
TYPE 'y' TO COMPUTE PERCENTILES:
------------------------------------------------------------------
  Input probabilities for which you wish to compute percentiles.
  The output is the probabilities in the column PROB and the
  corresponding percentiles in the column PERCNTLE.
------------------------------------------------------------------
y
DATA> .05 .25 .5 .75 .95
DATA> end

  ROW    PROB    PERCNTLE
    1    0.05    0.163020
    2    0.25    0.197911
    3    0.50    0.224230
    4    0.75    0.252104
    5    0.95    0.294626
```

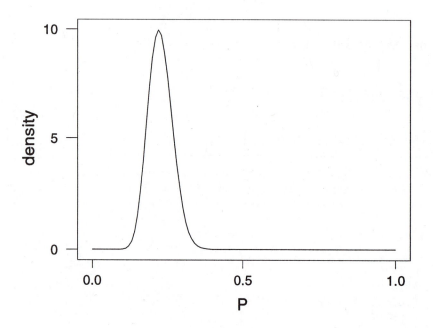

Figure 4.5: The beta (24.4, 83.6) posterior density for baseball example.

In addition, one can make updated predictions about the hitter's performance in future games using the macro 'p_beta_p'. This program is run to find the predictive distribution for 20 future at-bats. The output is presented here and Figure 4.6 displays the graph of the predictive probabilities. After the programs 'p_beta' and 'p_beta_p' have been run, we have learned

- the probability that the batter is under a .200 hitter is 27%,

- one good guess at the batting ability of the hitter is the median, or .5 percentile, .224, and

- if the player bats an additional 20 times, the most likely number of hits is four and the probability that he gets two or fewer hits is .009701 + .045765 + .107835 = .163.

```
MTB > exec 'p_beta_p'

INPUT VALUES OF BETA PARAMETERS A AND B:
DATA> 24.2 83.6

INPUT NUMBER OF TRIALS:
DATA> 20

INPUT RANGE (LOW AND HIGH VALUES) FOR NUMBER OF SUCCESSES:
DATA> 0 20

PREDICTIVE DISTRIBUTION OF NUMBER OF SUCCESSES:

ROW   SUCC      PRED
  1      0   0.009701
  2      1   0.045765
  3      2   0.107835
  4      3   0.168505
  5      4   0.195573
  6      5   0.178992
  7      6   0.133878
  8      7   0.083708
  9      8   0.044393
 10      9   0.020148
 11     10   0.007861
 12     11   0.002639
 13     12   0.000761
```

14	13	0.000187
15	14	0.000039
16	15	0.000007
17	16	0.000001
18	17	0.000000
19	18	0.000000
20	19	0.000000
21	20	0.000000

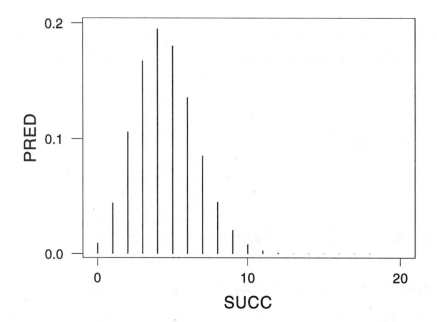

Figure 4.6: Predictive probabilities for beta(24.4, 83.6) density for baseball example.

4.3.5   A test for a proportion

> Minitab command to test for a proportion:
> **exec 'p_beta_t'**

The command 'p_beta_t' performs a Bayesian test of a binomial proportion. We are interested whether the proportion p is equal to a specific value. In the baseball example, we may wonder whether a particular player is a .400 hitter. That is, we wonder whether the batter's proportion of hits in the long run is equal to .4. When tossing a coin, we may be interested whether the coin is fair. If p represents the proportion of heads when the coin is tossed repeatedly, we question whether $p = .5$.

In this problem, there are two hypotheses or statements about the value of p. The hypothesis H is the statement that p is equal to the value of interest, p_0, and the alternative hypothesis K is that p is not equal to this value. To make a decision, we compute the posterior probabilities of the two hypotheses. Suppose that the prior probability of the hypothesis H that the proportion is equal to the specified value is q. Then the prior probability of the alternative hypothesis will be $1 - q$. Then, by Bayes' rule, the posterior probability of the hypothesis H is given by

$$P(\text{H given data}) = \frac{qBF}{qBF + 1 - q},$$

where BF is the Bayes factor. This Bayesian test statistic is the ratio of the probabilities of the data under the two hypotheses:

$$BF = \frac{P(\text{data given H})}{P(\text{data given K})}.$$

One can make a decision about the hypotheses on the basis of the Bayes factor or the posterior probabilities. The Bayes factor is a measure of the evidence in support of the hypothesis H contained in the data. If, for example, $BF = 10$, then the hypothesis $p = p_0$ is ten times more likely than the alternative hypothesis that $p \neq p_0$. The posterior probabilities of the hypotheses combine the evidence in the data with the opinions of the user before sampling. Suppose that one believes initially that $p = p_0$ with probability .25. If the Bayes factor $BF = 3$, then the posterior probability of p_0 is

$$P(H \text{ given data}) = \frac{.25 \times 3}{.25 \times 3 + 1 - .25} = .5.$$

Thus, after seeing the data, the user is indifferent between the two hypotheses.

A special type of prior distribution is developed for this testing situation. First, one gives an initial (prior) probability that the hypothesis H($p = p_0$) is true. Then one assigns a beta prior density with parameters a and b to the continuous proportion p if the proportion is not equal to p_0. Here one has to think of plausible alternative values for the proportion value. The beta density reflects the opinion of the user about the location of these alternative p values.

To illustrate the construction of this prior distribution, consider the earlier coin tossing example. Suppose one is interested in testing whether the coin is fair. So one is testing the hypothesis that $p = .5$. Suppose one is pretty certain that the coin is fair. Then the prior probability q would be assigned a large value such as .9. Next, one thinks about the probability of heads if the coin is not fair. Suppose that one thinks that the coin may have been fair when it was first made. But since this particular coin has been in circulation for many years, the two sides aren't quite balanced. So the probability of heads is likely to be in a small neighborhood of one-half. A beta density with parameters $a = 10$ and $b = 10$ that is symmetric about $p = .5$ may be a reasonable model for this prior opinion. A second person may think differently about this coin. She thinks that most coins are basically fair, but there exist some trick coins with two heads or two tails. In this case, her alternative prior would place much of its mass near the extreme values of 0 and 1. A beta density with values of a and b each smaller than 1 may be representative of these prior beliefs.

The program 'p_beta_t' will compute the Bayes factor and the posterior probabilities of the hypotheses for this situation. The output for this coin tossing example is shown here. One first inputs the value of p to be tested. If one wished to test whether a coin is fair, the person would type the value .5. Next, one inputs the probability of the null hypothesis H that the proportion is equal to the specified value. In this particular run, suppose that the person is indifferent about the coin being fair or unfair, and so this probability is set to .5. Next, one inputs the parameters a and b of the beta density, which represents the user's prior opinion if the alternative hypothesis K is true. The beta parameters $a = 10$ and $b = 10$ are used. This beta(10, 10) distribution says that, if the coin is not fair, values of p close to .5 are most likely. Last, one inputs the data, the number of observed successes (s) and failures (f). In this example, suppose that the coin is tossed 50 times with 22 heads and 28 tails. After printing the inputted values, the macro gives the Bayes factor BF in favor of the hypothesis that the proportion is equal to .5, the Bayes factor $1/BF$ in support of the alternative hypothesis, and the posterior probability that the hypothesis is true. Here the Bayes factor is equal to 1.45, which indicates that there is some support from the data for the "fair coin." In addition, the posterior probability of "fair coin" is .592, which is slightly larger than the prior probability of .5.

```
MTB > exec 'p_beta_t'

ENTER THE NULL HYPOTHESIS PROPORTION P0:
DATA> .5

ENTER THE PRIOR PROBABILITY OF P0:
DATA> .5

FOR THE ALTERNATIVE HYPOTHESIS THAT P = P0,
ENTER THE NUMBERS A AND B OF THE BETA(A, B) DISTRIBUTION:
DATA> 10 10

ENTER THE OBSERVED NUMBER OF SUCCESSES AND FAILURES:
DATA> 22 28

You have entered:

  ROW     P0   prob_H    a_b    data
    1    0.5     0.5      10      22
    2                     10      28

The Bayes factor in favor of the null hypothesis is:
BF_HK
  1.45372

The Bayes factor against the null hypothesis is:
BF_KH
  0.68789

The posterior probability of the null hypothesis is:
prob_H
  0.592456
```

4.3.6 Using a histogram prior

> Minitab command to learn about a proportion using a histogram prior:
> **exec 'p_hist_p'**

In this section, we consider an alternative form of prior density for a continuous proportion p. A general method of constructing a prior density for a proportion is first to divide the region of possible values into subintervals, and then assign probabilities to

all of the intervals. For example, again think of the baseball manager who is interested
in the proportion of hits of a particular major leaguer. To construct a prior density
for this proportion, he may think that the batting average p must lie between .2 and
.4, and then think about the likelihoods of p falling in the intervals (.2, .25), (.25, .3),
(.3, .35), (.35, .4). This assessment is similar to the specification of probabilities for a
proportion when a discrete set of values is of interest. If the manager is familiar with the
construction of probabilities in the discrete setting, then he would likely be comfortable
in specifying probabilities of intervals in this continuous setting.

One method of using this "histogram prior" is first to find the particular beta density that approximately matches this histogram and then use the program 'p_beta' to
summarize the beta posterior density. This method is attractive in that it is easy to
update a beta density. However, this matching can be difficult to perform, especially
when the shape of the histogram is not similar to a member of the beta family.

The program 'p_hist_p' provides a direct method of updating probabilities of intervals when one's prior is in the form of a histogram. We will illustrate this program for
this baseball example (see the output below). One first inputs the histogram prior by
entering the midpoints of the intervals and next entering the respective prior probabilities. In this case, the midpoints of the four intervals are .225, .275, .325, .375, and the
manager gives these intervals the respective prior probabilities .4, .4, .15, .05, reflecting
the belief that the hitter is likely average with a small probability of being a great hitter.

```
MTB > exec 'p_hist_p'

INPUT INTERVAL MIDPOINTS:
DATA> .225 .275 .325 .375
DATA> end

INPUT PRIOR PROBABILITIES OF INTERVALS:
DATA> .4 .4 .15 .05

INPUT NUMBER OF SIMULATED VALUES:
DATA> 1000

INPUT OBSERVED NUMBER OF SUCCESSES AND FAILURES:
DATA> 4 36

The prior and posterior probabilities of the intervals:
```

ROW	MIDS	LO	HI	PRIOR	POST
1	0.225	0.20	0.25	0.40	0.822267
2	0.275	0.25	0.30	0.40	0.164236
3	0.325	0.30	0.35	0.15	0.013097
4	0.375	0.35	0.40	0.05	0.000401

This program computes the posterior probabilities of the four intervals by means of simulation. A large number of proportion values are simulated from the prior histogram. The program asks for the number of simulated values to use — a large number over 500 should suffice for many applications. For each value of p simulated, the likelihood (the probability of observing a sequence of s successes and f failures) is computed. These likelihoods are converted into probabilities by dividing each likelihood by the sum. Then the probabilities are grouped into the intervals — the sum of probabilities in the first interval is the posterior probability that p is in this interval.

The last input is the number of successes (s) and failures (f) that are observed. In the table in the earlier output, the program gives the low and high values for each interval, and the associated prior probabilities, and the posterior probabilities of the intervals. Note that, after observing this data, 80% of one's probability falls in the first interval (.20, .25). One is quite confident that the player is a below-average hitter.

4.4 Exercises

1. (Berry[1], example 6.1.) In a segment of the TV show *The Odd Couple*, Felix claims to have ESP. To test his claim, Oscar proposes the following experiment. Oscar will select one from four large cards with different geometric figures and Felix will try to identify it. Let p denote the probability that Felix is correct in identifying the figure for a single card. Oscar believes that Felix has no ESP ability ($p = .25$), but there is a small chance that p is either larger or smaller than .25. After some thought, Oscar places the following prior distribution on p:

p	0	.125	.250	.375	.500	.625	.750	.875	1
$P(p)$	.001	.001	.950	.008	.008	.008	.008	.008	.008

---

[1]Berry, D. (1996), *Statistics: A Bayesian Perspective*, Belmont, CA.: Duxbury Press.

Suppose that the experiment is repeated ten times and Felix is correct six times and incorrect four times. Using the program 'p_disc', find the posterior probabilities of the above values of p. What is Oscar's posterior probability that Felix has no ability?

2. (Schmitt[2], exercise 3.2.7) A certain disease causes minks to lose patches of hair. A new drug has been developed to combat it, and preliminary testing indicates that its cure rate is in the vicinity of 2/3. Suppose that there are four alternative values for the cure rate .6, .65, .7, and .75, which are equally probable on the early evidence. Suppose that the drug is tested on twenty-nine animals; twenty are cured and nine are not.

   (a) What is the probability distribution of the cure rates after experimentation? (Use the program 'p_disc'.)

   (b) Suppose that twenty additional animals will be tested on the drug. Find the predictive distribution of the number that will be cured. What is the most likely number to be cured? (Use the program 'p_disc_p'.)

3. Suppose you are tossing a coin. You strongly believe (with probability .98) that the coin is fair. If p is the probability of a head, then $P(p = .5) = .98$. However, there is a small probability of .02 that you are tossing a trick coin that has two heads ($p = 1$). You now toss the coin and get 10 heads in a row. How has your opinion about the coin changed on the basis of this experiment?

4. (Berry[3], exercise 6.9) Someone claims she can tell the sex of an unborn child from the way it "rides" — high means boy and low means girl. She claims to be good but not perfect. Suppose you don't believe her and associate probability .7 that she's only guessing — the null hypothesis of no ability. Let p be the probability that she is correct and suppose that there are three other plausible values for p. The values of p and the prior probabilities are listed here:

---

[2]Schmitt, S. A. (1969),*Measuring Uncertainty: An Elementary Introduction to Bayesian Statistics*, New York: Addison-Wesley.

[3]Berry, D. (1996), *Statistics: A Bayesian Perspective*, Belmont, CA.: Duxbury Press.

p	.5	.625	.75	.875
$P(p)$	.7	.1	.1	.1

(a) Suppose you perform an experiment with ten expectant mothers to test the claim. Suppose that she is right on eight of the ten and not correct on the other two. Find your updated probabilities. (Use 'p_disc'.)

(b) If she predicts the sex of the children of five new expectant mothers, find the probability that she gets 0, 1, 2, 3, 4, 5 correct. (Use 'p_disc_p'.)

5. (Berry[4], exercise 6.11) Burrowing owls sometimes build their nests in holes that were dug (but have been since abandoned) by prairie dogs, coyotes, or badgers. They sometimes line the nests with cattle or horse dung. To explain this behavior, a biologist suggested that the owls use the dung to keep predators away. To test this theory, a second biologist observed lined and unlined owl nests in the Columbia River basin. We consider only the lined nest data here. Let p denote the probability that a owl nest with dung is raided. Suppose that the biologist's prior probabilities are as indicated here:

p	0	.1	.2	.3	.4	.5	.6	.7	.8	.9	1
$P(p)$	.08	.12	.08	.04	.12	.28	.12	.04	.04	.04	.04

(a) If the biologist observes twenty-five nests with dung and notes that only two nests were raided, find the posterior probabilities of the eleven models for p.

(b) If you are to find ten additional nests lined with dung, find the probability that none of them will be raided. Find the probability that at most two nests are raided.

6. (Berry[4], exercise 6.13) Researchers followed 127 adults under the age of 70 who had undergone an angioplasty procedure to widen at least one constricted heart artery. After eighteen months, twenty-eight of them had experienced serious reactions: severe chest pains, heart attacks or sudden death. Let p denote the probability that a patient under seventy who gets this procedure will experience a serious

---

[4]Berry, D. (1996), *Statistics: A Bayesian Perspective*, Belmont, CA.: Duxbury Press.

reaction. Consider the eleven possible p models 0, .1, .2, ..., 1 and assume that each model has the same probability.

One can specify this set of equally spaced values of p and the uniform probabilities by means of the Minitab commands

```
set 'p'
0:1/.1
end
let 'prior'=1/count('p')+0*'p'
```

(a) Given the data above, find the posterior probabilities of the eleven models.

(b) Consider ten additional patients under seventy who undergo an angioplasty procedure. Find the predictive distribution for the number who experience serious reactions.

7. (Berry[5], exercise 7.5) You plan to survey ten people's opinions concerning legalized abortion. Before the survey, you wish to construct a beta(a, b) prior density for the proportion of the population p who favor legalized abortion. You think that your probability that the first person is in favor is equal to .40. And if the first person you ask happens to be in favor, your probability the second is also in favor is .45. Of the ten people you ask, seven say they favor and three say they do not.

(a) Use the program 'beta_sel' to find the beta distribution(a, b) that matches the prior information. Use the program 'p_beta' to plot the beta prior density. In addition, find the prior probability that p is larger than .5, and find an interval that contains 90% of the prior probability.

(b) Compute the parameters of the posterior beta density (after you observe the responses of the ten people). Use 'p_beta' to plot this density and, as in part (a), find the updated probability that p is larger than .5 and find a 90% interval estimate. Describe how the posterior density is different from the prior density.

(c) Suppose that you are to survey ten new people. Using the program 'p_beta_p', find the predictive distribution for the number in this new sample that are in favor of legalized abortion.

---

[5]Berry, D. (1996), *Statistics: A Bayesian Perspective*, Belmont, CA.: Duxbury Press.

8. (Berry[6], exercise 7.6) Consider an experiment involving an ordinary thumbtack. Suppose the tack is dropped on a hard, flat surface and you observe whether it comes to rest with point up — call point up a "success." You should hold the tack two or three feet above the surface and drop it vertically — do not throw it.

   (a) Assuming a beta prior, use 'beta_sel' to assess the prior distribution for p, the probability of a success.

   (b) Plot your prior density using 'p_beta' and find a 90% probability interval for p.

   (c) Drop the tack a total of 10 times, keeping track of the number of successes. Find your posterior density. Plot the density and find an updated 90% interval estimate using the program 'p_beta'.

9. (Berry[6], exercise 7.28) A study reported on the long-term effects of exposure to low levels of lead in childhood. Researchers analyzed children's shed primary teeth for lead content. Of the children whose teeth had a lead content of more that 22.22 parts per million (ppm), twenty-two eventually graduated from high school and seven did not. Suppose your prior density for p, the proportion of all such children who will graduate from high school is beta(1, 1), and so your posterior density is beta(23, 8). Based on this information, of ten more children who are found to have lead content of more than 22.22 ppm, what is your predictive probability that nine or ten of them will graduate from high school? (Use the program 'p_beta_p'.)

10. Refer to exercise 8. Suppose that you are interested in testing the hypothesis that p, the proportion of times that the tack lands point up, is equal to .5.

    (a) Suppose that you initially assign the hypothesis a probability of .5. In addition, alternative values of the proportion p are possible with the following prior probabilities:

p	.1	.2	.3	.4	.5	.6	.7	.8	.9
$P(p)$	.01	.02	.08	.14	.50	.14	.08	.02	.01

    Using the data that you collected in exercise 8 and the program 'p_disc', find the posterior probability of $p = .5$.

---

[6]Berry, D. (1996), *Statistics: A Bayesian Perspective*, Belmont, CA.: Duxbury Press.

(b) Consider an alternative method of testing this hypothesis using a continuous prior. As in part (a), you initially assign this hypothesis a probability of .5. If the hypothesis is not true, suppose that you think that values of p close to .5 are most likely and so assign this probability a beta(5, 5) prior. Use the program 'p_beta_t' to compute the posterior probability of $p = .5$.

(c) Compare the results in parts (a) and (b).

(d) How sensitive is the posterior probability computed in part (b) to the assumption of a beta(5, 5) prior? Rerun 'p_beta_t' using a beta(1, 1) (uniform) prior on p when the hypothesis is not true. Compare your answer to the one obtained in part (b).

11. Consider the following experiment. Hold a penny on edge on a flat hard surface, and spin it with your fingers. Let p denote the probability that it lands heads. To estimate this probability, we will use a histogram to model our prior beliefs about p. Divide the interval [0,1] into the 10 subintervals [0,.1], [.1,.2], ..., [.9,1], and specify probabilities that p is in each interval. Next spin the penny twenty times and count the number of successes (heads) and failures (tails). To update the probabilities that you assigned to the intervals, run 'p_hist_p'. The input to this program are the midpoints of the intervals (.05, .15, ..., .95), the prior probabilities, and the data (number of successes and failures). How have the interval probabilities changed on the basis of your data?

Chapter 5

Comparing Two Proportions

5.1 Introduction

This chapter describes Minitab programs to compare two population proportions. Suppose that two populations are of interest. Each population is divided into two groups, successes and failures, and you are interested in comparing the proportion of successes in population 1, p_1, with the proportion of successes in population 2, p_2. To illustrate this situation, consider example 8.1 from Berry[1] where one is interested whether people who carry the sickle-cell gene are more resistant to malaria than those people who don't carry the gene. Here there are two populations — those who have the sickle-cell gene and those people who don't. An experiment was conducted where subjects were injected with malaria parasites. Let p_1 denote the proportion of the sickle-cell population who would develop malaria when exposed and p_2 the corresponding proportion for the non-sickle-cell population.

To learn about the proportions, independent samples from the two populations are taken. Suppose that one observes s_1 successes and f_1 failures in the sample from population 1 and s_2 successes and f_2 failures in the second sample. In the sickle-cell study, only two of the fifteen sickle-cell subjects developed malaria and fourteen of the fifteen non-sickle-cell subjects developed the disease.

There are two basic inference questions that will be considered in this setting. First, is there sufficient evidence from the data that the two proportions are different? Second,

---

[1]Berry, D. (1996), *Statistics: A Bayesian Perspective*, Belmont, CA.: Duxbury Press.

if the proportions appear to be different, one may be interested in learning about the magnitude of the difference. In this chapter, we'll focus on the difference of proportions $d = p_2 - p_1$ and describe programs that will help in learning about the size of d.

As in Chapter 4, the programs will illustrate two general methods for modeling prior information about the two proportions. The *discrete* approach, described in Section 5.2, selects a set of possible values for each proportion and then constructs a prior distribution on a grid of possible pairs of proportion values. Section 5.3 discusses the *continuous* approach to modeling where each proportion is assumed to be continuous-valued. In this case, a bivariate density function for (p_1, p_2) is used to represent prior beliefs about the locations of the proportions.

5.2   Using discrete models

5.2.1   Constructing a prior

To illustrate the application of discrete models for two proportions, consider the malaria study discussed in the introduction. To learn about the two population proportions, one first constructs a prior distribution. For each proportion, a set of plausible values needs to specified. For this example, suppose that little is known about the proportion who would get malaria when exposed for either the sickle-cell or non-sickle-cell populations. Since proportion values from 0 to 1 are all possible, we'll let each proportion take on the five equally spaced values 0, .25, .5, .75, 1.

Since the values of both proportions are unknown, a model is an ordered pair (p_1, p_2). We can represent the collection of all models by the following matrix, where the possible values of the first proportion are listed down the rows and the values for the second proportion are listed across the columns.

	p_2				
p_1	0	.25	.5	.75	1
1					
.75					
.5					
.25					
0					

The next step is to assign probabilities to the different models. In this case, this means that one needs to assign a number to each cell of the 5-by-5 table that reflects the plausibility of that particular pair of proportion values. This can be a difficult task, so the Minitab programs are set up to use particular default choices for the probabilities that reflect different prior opinions.

In the case where one has little prior information about either proportion, then one can use a *uniform* distribution where each model is assigned the same probability. In this example, each of the twenty-five possible pairs of proportions is given a probability of $1/25 = .04$.

	p_2				
p_1	0	.25	.5	.75	1
1	.04	.04	.04	.04	.04
.75	.04	.04	.04	.04	.04
.5	.04	.04	.04	.04	.04
.25	.04	.04	.04	.04	.04
0	.04	.04	.04	.04	.04

The program 'pp_disc', described in Section 5.2.2, will use such a uniform prior distribution.

Another possibility is that the user believes, with high probability, that the two proportions are equal. In our example, suppose that she believes that there is a good chance that the sickle-cell and non-sickle-cell populations have the same risk of getting malaria when affected. If she thinks that the events "have the same risk" and "have different risks" have the same probability, then the proportions p_1 and p_2 would be equal with probability .5. One can model this belief by assigning larger probabilities on the *diagonal* models $(p_1, p_2) = (0, 0), \ldots, (1, 1)$ where the two proportions are equal. If each diagonal model is assigned the same probability and each nondiagonal model (where the proportions are unequal) is given the same probability, then one obtains the following set of probabilities:

			p_2		
p_1	0	.25	.5	.75	1
1	.025	.025	.025	.025	.1
.75	.025	.025	.025	.1	.025
.5	.025	.025	.1	.025	.025
.25	.025	.1	.025	.025	.025
0	.1	.025	.025	.025	.025

This particular form of prior distribution will be called a "testing prior," since it can be used to test the hypothesis that the two proportions are equal. The program 'pp_disct', described in Section 5.2.3, will construct this type of prior distribution.

In other situations, one may have more information about the locations of the two proportions. One alternative method of constructing a prior distribution is to first assign probabilities to each proportion separately and then, by assuming independence between p_1 and p_2, multiply the two marginal probability distributions to find joint prior probabilities. For example, suppose that the user believes before looking at any data that the sickle-cell gene is effective in protecting a person from malaria. Moreover, if he doesn't have the gene, then he has a high probability of getting malaria. He then assigns the probabilities $\{.4, .5, .1, 0, 0\}$ to the values $\{0, .25, .5, .75, 1\}$ for p_1 and the probabilities $\{0, 0, .2, .4, .4\}$ to the same proportion values for p_2. Last, he believes that his opinions about the sickle-cell proportion are unrelated to his beliefs about the non-sickle-cell proportion. Then, for example, the probability that $p_1 = .25$ and $p_2 = .5$ can be obtained by multiplying the probability that $p_1 = .25$ and the probability of $p_2 = .5$. If the probabilities of all pairs of models are computed in a similar fashion, the following table is obtained:

			p_2		
p_1	0	.25	.5	.75	1
1	0	0	0	0	0
.75	0	0	0	0	0
.5	0	0	.02	.04	.04
.25	0	0	.10	.20	.20
0	0	0	.08	.16	.16

The program 'pp_discm', discussed in Section 5.2.4, can be used to input an arbitrary table of prior probabilities such as this one.

5.2.2 Computing the posterior distribution

After one has constructed the table of probabilities, then it is easy in principle to update these probabilities after data have been observed. Suppose that one observes s_1 successes and f_1 failures in the first sample and s_2 successes and f_2 failures in the second sample. Then, for a particular set of proportion values for p_1 and p_2, the likelihood or probability of observing this sample result is given by

$$\text{LIKE} = p_1^{s_1}(1 - p_1)^{f_1} p_2^{s_2}(1 - p_2)^{f_2}.$$

These likelihoods are computed for all of the proportion pairs in the table. Posterior probabilities are found by multiplying these likelihoods by the corresponding prior probabilities to obtain products, and then dividing these products by the total sum of the products to get updated probabilities.

One can learn about the difference of proportions $d = p_2 - p_1$ from the table of posterior probabilities. For the malaria/sickle-cell example, if one uses a uniform prior and the observed data are $(s_1, f_1, s_2, f_2) = (2, 13, 14, 1)$, then the posterior probabilities are given by the following table:

			p_2		
p_1	0	.25	.5	.75	1
1	0	0	0	0	0
.75	0	0	0	0	0
.5	0	0	0	.020	0
.25	0	0	.007	.973	0
0	0	0	0	0	0

The probability that the difference in proportions d is equal to 0 is found by summing the probabilities over the diagonal cells where $p_1 = p_2$. Similarly, the probability that d is equal to .25 is the sum of the probabilities that $(p_1, p_2) = (0, .25), (.25, .5), (.5, .75),$ $(.75, 1)$. The posterior probabilities for all possible values of d are summarized in this table:

$d = p_2 - p_1$	-1	$-.75$	$-.5$	$-.25$	0	.25	.50	.75	1
PROB	0	0	0	0	0	.027	.973	0	0

In this case, most of the probability for d is concentrated on the single value .5. In the examples to follow, more detailed information about the difference in proportions is obtained by using a finer grid for the two proportions.

5.2.3   Using a uniform prior

> Minitab command to learn about two proportions using discrete uniform prior:
> **exec 'pp_disc'**

The program 'pp_disc' constructs a uniform prior distribution over an equally spaced grid of values for each proportion. In addition, the program computes posterior probabilities for the proportions and the difference in proportions d. The output of this program for the malaria/sickle-cell example is shown. The user is asked to input the smallest and largest values for each proportion and the number of models. Here each proportion will take on eleven equally spaced values from 0 to 1. Last, she inputs the data — the number of successes and failures for each of the two samples.

The posterior probabilities of the two proportions are displayed in a table format. The values of the first proportion p_1 correspond to the rows of the table, and the values for p_2 correspond to the columns. This table of probabilities is represented by a scatterplot in Figure 5.1, where the location of a plotted plot corresponds to a particular (p_1, p_2) model, and the symbol of the plotted point corresponds to the size of the corresponding probability. The probability values that exceed one-half of the maximum probability are darkest, the values between one-tenth and one-half of the maximum probability are less dark, and small probability values correspond to a dot. In this example, note that the probability distribution is clustered toward small p_1 values and large p_2 values.

```
MTB > exec 'pp_disc'

 FOR EACH P DISTRIBUTION:
 ------------------------
 INPUT LO AND HI VALUES:
DATA> 0 1

 INPUT NUMBER OF MODELS:
DATA> 11

INPUT OBSERVED NUMBER OF SUCCESSES AND FAILURES IN FIRST SAMPLE:
DATA> 2 13

INPUT OBSERVED NUMBER OF SUCCESSES AND FAILURES IN SECOND SAMPLE:
DATA> 14 1
```

Posterior distribution of P1 and P2:
(Rows and columns are expressed in percentage format.)

ROWS: PER_1   COLUMNS: PER_2

	0	10	20	30	40	50	60	70
0	0.000000	0.000000	0.000000	0.000000	0.000000	0.000000	0.000000	0.000000
10	0.000000	0.000000	0.000000	0.000000	0.000021	0.000389	0.003996	0.025940
20	0.000000	0.000000	0.000000	0.000000	0.000018	0.000337	0.003457	0.022441
30	0.000000	0.000000	0.000000	0.000000	0.000007	0.000133	0.001371	0.008899
40	0.000000	0.000000	0.000000	0.000000	0.000002	0.000032	0.000329	0.002133
50	0.000000	0.000000	0.000000	0.000000	0.000000	0.000005	0.000048	0.000311
60	0.000000	0.000000	0.000000	0.000000	0.000000	0.000000	0.000004	0.000025
70	0.000000	0.000000	0.000000	0.000000	0.000000	0.000000	0.000000	0.000001
80	0.000000	0.000000	0.000000	0.000000	0.000000	0.000000	0.000000	0.000000
90	0.000000	0.000000	0.000000	0.000000	0.000000	0.000000	0.000000	0.000000
100	0.000000	0.000000	0.000000	0.000000	0.000000	0.000000	0.000000	0.000000

	80	90	100
0	0.000000	0.000000	0.000000
10	0.112142	0.291657	0.000000
20	0.097016	0.252319	0.000000
30	0.038471	0.100055	0.000000
40	0.009219	0.023978	0.000000
50	0.001346	0.003502	0.000000
60	0.000107	0.000277	0.000000
70	0.000003	0.000009	0.000000
80	0.000000	0.000000	0.000000
90	0.000000	0.000000	0.000000
100	0.000000	0.000000	0.000000

TYPE 'Y' AND RETURN TO SEE A GRAPH OF THE POSTERIOR DISTRIBUTION:
y

TYPE 'Y' AND RETURN TO SEE A TABLE OF THE POSTERIOR DISTRIBUTION
OF THE DIFFERENCE IN PROBABILITIES P2-P1:
y

Row	DIFF	P_DIFF
1	-1.0	0.000000
2	-0.9	0.000000
3	-0.8	0.000000
4	-0.7	0.000000
5	-0.6	0.000000

```
 6   -0.5   0.000000
 7   -0.4   0.000000
 8   -0.3   0.000000
 9   -0.2   0.000000
10   -0.1   0.000001
11    0.0   0.000111
12    0.1   0.000116
13    0.2   0.000907
14    0.3   0.005484
15    0.4   0.025464
16    0.5   0.088877
17    0.6   0.222989
18    0.7   0.364424
19    0.8   0.291628
20    0.9   0.000000
21    1.0   0.000000
```

The program also prints the possible values and corresponding probabilities for the difference in proportions $d = p_2 - p_1$. By inspection of this distribution of probabilities, we see that over 95% of the probability is concentrated on the values $d = .5, .6, .7, .8$.

The prior and posterior probabilities are stored in columns of the Minitab worksheet. The column 'P1' contains values of the proportion p_1, and the column 'P2' contains values of p_2. The uniform prior probabilities are stored in the column 'PRIOR' and the posterior probabilities in the column 'POST'. One can perform computations on these columns to find any probability of interest. To illustrate, suppose that one is interested in the posterior probability that the sickle-cell probability p_1 is smaller than the non-sickle-cell probability p_2. This can be found using the Minitab commands

```
MTB > let k1=sum(('p1'<'p2')*'post')
MTB > print k1
K1        0.999988
```

The probability that the second proportion is larger is essentially one.

The posterior probabilities for the difference in proportions d are also stored in the worksheet. The values of d are stored in the column 'DIFF' and the associated probabilities in the column 'P_DIFF'. One can use the program 'disc_sum' (described in Section 4.2.2) to summarize this probability distribution. For example, suppose one was interested in the probability that the difference in proportions is less than or equal to .7. Run the program 'disc_sum' and input the numbers of the columns that contain

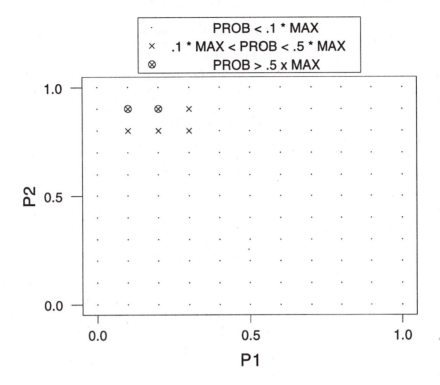

Figure 5.1: Posterior probabilities of two proportions using a uniform prior for malaria example.

the distribution for d (the column 'DIFF' is the same as C55 and 'P_DIFF' the same as C56). By inputting the number .7 as a value for which you wish to compute cumulative probabilities, the program will output $P(d \leq .7) = .708$.

5.2.4 A test of the equality of two proportions

Minitab command to test the equality of two proportions using discrete models:
exec 'pp_disct'

For the malaria/sickle-cell example, suppose one is interested whether this data provide sufficient evidence that the proportions p_1 and p_2 are not equal. A Bayesian test of the hypotheses H: $p_1 = p_2$, K:$p_1 \neq p_2$ can be developed by use of a special type of discrete prior distribution. The program 'pp_disct' will construct this "testing" form of prior distribution and compute posterior probabilities for the two proportions.

A run of the program 'pp_disct' is shown for the malaria/sickle-cell data. As in the previous section, one first chooses the grid of possible proportion models. In this example, we select eleven possible models for each proportion between 0 and 1. Next, one inputs the prior probability of the hypothesis H that the two proportions are equal. In this case, we believe that the hypotheses of equality and non-equality are equally likely, and so we assign H a prior probability of .5. Each pair of proportion values within each hypothesis is assigned the same prior probability. In this case there are eleven possible diagonal models for (p_1, p_2) where $p_1 = p_2$. Since we want the total probability of equality to be equal to .5, each pair of diagonal models is assigned the probability of $.5/11 = .0454$. Likewise, since there are 110 models for the proportions where the proportions are unequal, each model is assigned the probability $.5/110 = .0045$.

```
MTB > exec 'pp_disct'

 FOR EACH P DISTRIBUTION:
 -------------------------
 INPUT LO AND HI VALUES:
 DATA> 0 1

 INPUT NUMBER OF MODELS:
 DATA> 11

 INPUT PROBABILITY THAT P1=P2:
 DATA> .5

INPUT OBSERVED NUMBER OF SUCCESSES AND FAILURES IN FIRST SAMPLE:
DATA> 2 13

INPUT OBSERVED NUMBER OF SUCCESSES AND FAILURES IN SECOND SAMPLE:
DATA> 14 1
```

Posterior distribution of P1 and P2:
(Rows and columns are expressed in percentage format.)

ROWS: PER_1   COLUMNS: PER_2

	0	10	20	30	40	50	60	70
0	0.000000	0.000000	0.000000	0.000000	0.000000	0.000000	0.000000	0.000000
10	0.000000	0.000000	0.000000	0.000000	0.000021	0.000389	0.003996	0.025937
20	0.000000	0.000000	0.000000	0.000000	0.000018	0.000337	0.003457	0.022439
30	0.000000	0.000000	0.000000	0.000001	0.000007	0.000133	0.001371	0.008898
40	0.000000	0.000000	0.000000	0.000000	0.000017	0.000032	0.000329	0.002132
50	0.000000	0.000000	0.000000	0.000000	0.000000	0.000047	0.000048	0.000311
60	0.000000	0.000000	0.000000	0.000000	0.000000	0.000000	0.000038	0.000025
70	0.000000	0.000000	0.000000	0.000000	0.000000	0.000000	0.000000	0.000008
80	0.000000	0.000000	0.000000	0.000000	0.000000	0.000000	0.000000	0.000000
90	0.000000	0.000000	0.000000	0.000000	0.000000	0.000000	0.000000	0.000000
100	0.000000	0.000000	0.000000	0.000000	0.000000	0.000000	0.000000	0.000000

	80	90	100
0	0.000000	0.000000	0.000000
10	0.112130	0.291628	0.000000
20	0.097007	0.252294	0.000000
30	0.038467	0.100045	0.000000
40	0.009218	0.023975	0.000000
50	0.001346	0.003501	0.000000
60	0.000107	0.000277	0.000000
70	0.000003	0.000009	0.000000
80	0.000000	0.000000	0.000000
90	0.000000	0.000000	0.000000
100	0.000000	0.000000	0.000000

TYPE 'Y' AND RETURN TO SEE A GRAPH OF THE POSTERIOR DISTRIBUTION:
y

TYPE 'Y' AND RETURN TO SEE A TABLE OF THE POSTERIOR DISTRIBUTION
OF THE DIFFERENCE IN PROBABILITIES P2-P1:
y

Row	DIFF	P_DIFF
1	-1.0	0.000000
2	-0.9	0.000000
3	-0.8	0.000000
4	-0.7	0.000000
5	-0.6	0.000000
6	-0.5	0.000000

```
 7    -0.4   0.000000
 8    -0.3   0.000000
 9    -0.2   0.000000
10    -0.1   0.000001
11     0.0   0.000111
12     0.1   0.000116
13     0.2   0.000907
14     0.3   0.005484
15     0.4   0.025464
16     0.5   0.088877
17     0.6   0.222989
18     0.7   0.364424
19     0.8   0.291628
20     0.9   0.000000
21     1.0   0.000000
```

As before, the values of the proportions p_1 and p_2 are stored in the Minitab columns 'P1' and 'P2' and the respective prior probabilities in the column 'PRIOR'. After one inputs the data (the number of successes and failures for each sample), the probabilities are updated. As in the earlier estimation problem, the posterior probabilities are found by multiplying the prior probabilities by the likelihood values and dividing each product by the total sum of the products. The table of posterior probabilities is displayed, and this distribution is graphed by a scatterplot. (This scatterplot is not shown here but resembles the one displayed in Section 5.2.2.) The values of the posterior probabilities are stored in the column 'POST'.

One tests the hypothesis that the proportions are equal by computing its posterior probability and comparing this value to the prior probability. The posterior probability that $p_1 = p_2$ is equivalent to the probability that the difference $d = 0$. In the output, we see that this probability is equal to .0000111, which is much smaller than the prior probability of .5. The evidence provided in the data in support of the hypothesis H is measured by the Bayes factor, which is the ratio of the posterior odds to the prior odds of the hypothesis. The prior odds of the hypothesis of equality of proportions is given by

$$O(H) = P(H)/P(K) = .5/.5 = 1,$$

and the corresponding posterior odds is

$$O(H|\text{data}) = P(H|\text{data})/P(K|\text{data}) = .000111/.999889 = .000111.$$

The Bayes factor in support of equality of proportions is

$$BF = \frac{O(\text{H}|\text{data})}{O(\text{H})} = \frac{.000111}{1} = .000111.$$

This value is very small, indicating strong support for the alternative hypothesis that the proportions are unequal.

5.2.5 Using a subjective prior

Minitab command to learn about two proportions using a subjective prior:
exec 'pp_discm'

The programs 'pp_disc'and 'pp_disct' are useful for learning about two proportions when there is relatively little prior information about their locations. The program 'pp_discm' is useful for updating probabilities for an arbitrary subjective prior distribution on a grid of proportion values.

To illustrate the application of this program for the sickle-cell/malaria example, suppose that the experimenter suspects, before seeing any data, that the sickle-cell gene is very effective in preventing malaria. Moreover, from previous experiments, she believes that people have a high probability of contracting malaria, when affected, if they don't carry the sickle-cell gene. To construct a prior distribution, she first considers prior probabilities for the proportion p_1 of sickle-cell carriers who get malaria. Values of the proportion in the set $\{.1, .2, \ldots, .9\}$ are believed possible, and she assigns the probabilities in the following table. This distribution says that the most likely value of p_1 is .1, and values greater than .6 are not possible.

p_1	.1	.2	.3	.4	.5	.6	.7	.8	.9
PROB	.4	.3	.15	.05	.05	.05	0	0	0

In a similar fashion, a prior distribution for the proportion of non-sickle-cell carriers who contract malaria, p_2 is found. The same set of proportion values is used, but the assigned probabilities are a mirror image of the ones just given. This distribution says that at least 40% of the non-sickle-cell group will get malaria, and the most probable value is 90%.

p_2	.1	.2	.3	.4	.5	.6	.7	.8	.9
PROB	0	0	0	.05	.05	.05	.15	.3	.4

To construct a joint distribution for the two proportions, the user makes the important assumption that her opinions about p_1 and p_2 are independent. In other words, her feelings about the proportion of non-sickle-cell people getting malaria won't change if she learns more about the proportion of sickle-cell people that get malaria. In this case, one can obtain probabilities for the proportion pairs (p_1, p_2) by multiplying the marginal distributions. For example, the joint probability of $p_1 = .1$ and $p_2 = .9$ is found by multiplying the probability of $p_1 = .1$ by the probability of $p_2 = .9$.

The proportion values and the prior probability matrix of the two proportions are placed in the Minitab worksheet using a number of set commands. In the output here, the values of the proportions p_1 and p_2 are placed in columns C1 and C2. Columns C3 through C11 contain the probability matrix. The first column C3 contains the probabilities for all values of p_1 for the first value of p_2, the second column C4 contains the probabilities for all values of p_1 for the second value of p_2, and so on.

```
MTB > prin c1 c2

Row      C1      C2
 1      0.1     0.1
 2      0.2     0.2
 3      0.3     0.3
 4      0.4     0.4
 5      0.5     0.5
 6      0.6     0.6
 7      0.7     0.7
 8      0.8     0.8
 9      0.9     0.9


MTB > prin c3-c11

Row   C3   C4   C5      C6       C7       C8       C9      C10     C11
 1    0    0    0    0.0200   0.0200   0.0200   0.0600   0.120   0.16
 2    0    0    0    0.0150   0.0150   0.0150   0.0450   0.090   0.12
 3    0    0    0    0.0075   0.0075   0.0075   0.0225   0.045   0.06
 4    0    0    0    0.0025   0.0025   0.0025   0.0075   0.015   0.02
 5    0    0    0    0.0025   0.0025   0.0025   0.0075   0.015   0.02
 6    0    0    0    0.0025   0.0025   0.0025   0.0075   0.015   0.02
 7    0    0    0    0.0000   0.0000   0.0000   0.0000   0.000   0.00
 8    0    0    0    0.0000   0.0000   0.0000   0.0000   0.000   0.00
 9    0    0    0    0.0000   0.0000   0.0000   0.0000   0.000   0.00
```

To update these probabilities, one runs the program 'pp_discm'. One inputs the prior distribution by entering the numbers of the columns that contain the values of the first and second proportion. The values of the first proportion are in the column C1, so the number 1 is typed. Similarly, 2 is the number of the column that contains values of p_2. Next, one inputs the number of the first column that contains the probability matrix. Here the probability matrix begins in column C3, and therefore the number 3 is typed. After one inputs the binomial data, the program computes and displays the posterior probabilities in the same format as the programs 'pp_disc' and 'pp_disct'. To compare the two proportions, the program outputs the posterior probabilities for the difference in proportions $d = p_2 - p_1$. The probability distribution for the difference in proportions is stored in the columns 'DIFF' and 'P_DIFF'.

```
MTB > exec 'pp_discm'

 INPUT THE NUMBER OF THE COLUMN WHICH CONTAINS THE P1 VALUES:
DATA> 1

 INPUT THE NUMBER OF THE COLUMN WHICH CONTAINS THE P2 VALUES:
DATA> 2

 INPUT THE NUMBER OF THE FIRST COLUMN WHICH CONTAINS THE PROBABILITIES:
DATA> 3

INPUT OBSERVED NUMBER OF SUCCESSES AND FAILURES IN FIRST SAMPLE:
DATA> 2 13

INPUT OBSERVED NUMBER OF SUCCESSES AND FAILURES IN SECOND SAMPLE:
DATA> 14 1

Posterior distribution of P1 and P2:
(Rows and columns are expressed in percentage format.)

 ROWS: PER_1     COLUMNS: PER_2

            10        20        30        40        50        60        70        80
10 0.000000 0.000000 0.000000 0.000004 0.000070 0.000723 0.014082 0.121756
20 0.000000 0.000000 0.000000 0.000002 0.000046 0.000469 0.009137 0.079000
30 0.000000 0.000000 0.000000 0.000000 0.000009 0.000093 0.001812 0.015663
40 0.000000 0.000000 0.000000 0.000000 0.000001 0.000007 0.000145 0.001251
50 0.000000 0.000000 0.000000 0.000000 0.000000 0.000001 0.000021 0.000183
60 0.000000 0.000000 0.000000 0.000000 0.000000 0.000000 0.000002 0.000014
```

```
70 0.000000 0.000000 0.000000 0.000000 0.000000 0.000000 0.000000 0.000000
80 0.000000 0.000000 0.000000 0.000000 0.000000 0.000000 0.000000 0.000000
90 0.000000 0.000000 0.000000 0.000000 0.000000 0.000000 0.000000 0.000000


            90
10 0.422216
20 0.273952
30 0.054316
40 0.004339
50 0.000634
60 0.000050
70 0.000000
80 0.000000
90 0.000000


   CELL CONTENTS --
              POST:DATA

TYPE 'Y' AND RETURN TO SEE A GRAPH OF THE POSTERIOR DISTRIBUTION:
y
TYPE 'Y' AND RETURN TO SEE A TABLE OF THE POSTERIOR DISTRIBUTION
OF THE DIFFERENCE IN PROBABILITIES P2-P1:
y

Row   DIFF     P_DIFF
   1   -0.8   0.000000
   2   -0.7   0.000000
   3   -0.6   0.000000
   4   -0.5   0.000000
   5   -0.4   0.000000
   6   -0.3   0.000000
   7   -0.2   0.000000
   8   -0.1   0.000000
   9    0.0   0.000000
  10    0.1   0.000004
  11    0.2   0.000054
  12    0.3   0.000520
  13    0.4   0.004236
  14    0.5   0.029862
  15    0.6   0.147399
  16    0.7   0.395708
  17    0.8   0.422216
```

It is interesting to compare the posterior probabilities using this subjective prior with the ones obtained with the uniform prior in Section 5.2.1. The posterior probabilities for the difference in proportions d are concentrated on larger values for the subjective prior analysis. For example, the probability that d is .7 or higher is .82 for the subjective prior analysis and .66 for the uniform prior analysis. This is reasonable, since the subjective analysis had a prior belief that the sickle-cell gene was effective in preventing malaria and so this person should have an especially strong opinion of this effect after seeing the data.

5.3 Using continuous models

5.3.1 Introduction

As in the previous section, we are interested in learning about two proportions p_1 and p_2. We are interested in learning about the size of the difference in proportions and in testing the hypothesis that the proportions are equal. What is new in this section is that continuous prior models are used for the proportions.

Suppose that each proportion is believed to be continuous valued from 0 to 1. Prior probabilities about the two proportions are described by a bivariate density function defined on the unit square. Although one can define different families of bivariate density functions, they generally are not very useful in practice in representing prior opinion. It is much harder for a user to specify a bivariate distribution for two proportions than to specify a density function for a single proportion. One reason for this difficulty is that the dependence between two proportions is hard to elicit. It is harder to think about the correlation between two proportions than to think of the general location of one proportion.

Since a bivariate prior density function is difficult to assess, the programs described in the section simplify the assessment process by making particular assumptions about the form of the prior density. Suppose that one's beliefs about the proportions p_1 and p_2 are independent. That is, suppose that one's opinion about one proportion, say p_1, is not affected by information that the user may obtain about the second proportion. Then one can construct a joint prior density for the pair of proportions (p_1, p_2) by separately assigning prior densities to p_1 and p_2. Suppose that a beta density with parameters a_1 and b_1 is a suitable model for reflecting one's information about p_1 and a beta density with parameters a_2 and b_2 represents prior opinion about p_2. (The programs 'beta_sel', 'p_beta', and 'p_beta_p' described in Chapter 4 can be used in the construction of these two beta prior densities.) If the data are the observed successes and failures in the two binomial experiments (s_1, f_1, s_2, f_2), then the posterior distributions for p_1 and p_2 will also be independent, with p_1 distributed beta$(a_1 + s_1, b_1 + f_1)$ and p_2 distributed beta$(a_2 + s_2, b_2 + f_2)$. The program 'pp_beta' is designed to summarize the posterior distribution of the difference in proportions $d = p_2 - p_1$ when the prior density is this independent beta form.

The program 'pp_beta' is helpful in learning about the general location of the difference in proportions. A different inference problem is testing the equality of two proportions. The program 'pp_bet_t' constructs a Bayesian test of the hypothesis H: $p_1 = p_2$ against the alternative hypothesis K: $p_1 \neq p_2$. For this program, two prior distributions must be specified: one on the region where the proportions are equal, and a second on the region where the proportions are unequal. This program uses beta densities for each of these two distributions. The program outputs two quantities: the posterior probability that the hypothesis H is true and the Bayes factor, which measures the evidence in the data in support of the hypothesis H .

Both of the programs 'pp_beta' and 'pp_bet_t' require the user to specify the location of one or more proportions by use of beta densities. It may be difficult to know before sampling the location of each proportion, but one may think that the two proportions are similar in size. One's prior beliefs about the proportions can be exchangeable, in that one's beliefs about one proportion are the same as the beliefs about the second proportion. The program 'pp_exch' can be used to model this belief in exchangeability. One inputs a parameter that indicates the degree of similarity of the two proportions and the program finds the posterior distribution of the two proportions. This program provides a method for combining the data from two different but related binomial samples.

5.3.2 Estimating the difference between two proportions

> Minitab command to learn about two proportions using continuous models:
> **exec 'pp_beta'**

Let's return to example 1.3 from Berry[2] that was discussed in Section 5.1. One is interested in estimating the improvement $d = p_2 - p_1$ in resistance, where p_1 is the

---

[2]Berry, D. (1996), *Statistics: A Bayesian Perspective*, Belmont, CA.: Duxbury Press.

proportion of sickle-cell gene carriers receiving malaria and p_2 is the proportion of noncarriers receiving malaria. Suppose that all possible values of the probabilities p_1 and p_2 are possible, and one assigns uniform prior distributions on the intervals $(0, 1)$ to reflect relative ignorance about the locations of these probabilities. These uniform prior distributions are a special case of beta distributions, where the parameter values are $a = 1$ and $b = 1$. If one had significant prior information about the probabilities, then the program 'beta_sel' (discussed in Chapter 4) could be run for p_1 and for p_2 to help in assessing suitable beta prior distributions.

For this example, the data are the number of successes and failures in the two samples $(s_1, f_1, s_2, f_2) = (2, 13, 14, 1)$. Then the posterior densities for p_1 and p_2 are independent, with p_1 distributed beta$(2+1, 13+1)$ and p_2 distributed beta$(14+1, 1+1)$.

In the program 'pp_beta', the posterior density of the difference in proportions $d = p_2 - p_1$ is obtained by means of simulation. By means of a random number generator, the program simulates a fixed number of values from a beta$(3, 14)$ distribution for p_1. Similarly, the same number of values is simulated from a beta$(15, 2)$ distribution for p_2. These two sets of random numbers are generated separately, since the two beta distributions are independent. Simulated values of the difference in proportions d are then found by matching the two sets of numbers and subtracting the random pairs. For example, if .314 was the first simulated value for p_1 and .521 the first simulated value for p_2, then the first simulated value of the difference d would be $.521 - .314 = .207$.

The output of the program 'pp_beta' is displayed here. One inputs the parameters of the two independent beta distributions for p_1 and p_2 and the number of values of the difference in proportions d to simulate. To get a good description of the posterior distribution of d, at least 500 simulated values should be used; our example uses 1000 values. The simulated values of p_1 and p_2 are stored in the Minitab columns 'P1' and 'P2' and the simulated values for the difference in proportions $d = p_2 - p_1$ are stored in the column 'P2-P1'.

This program gives different graphical summaries of the simulated values. The simulated values of the proportions p_1 and p_2 are displayed using a scatterplot. A dotplot is used to show the 1000 simulated values of the difference in proportions d. Looking at the figure, we see for the example that most of the posterior probability is concentrated in a small square where $0 < p_1 < .3$ and $.7 < p_2 < 1$. In addition, the dotplot of the simulated values of $d = p_2 - p_1$ indicates that the most likely value of the difference is about .7, and most of the difference values fall between .45 and .90.

The program can also compute probabilities regarding the difference in proportions. Suppose, for example, that one is interested in computing the probability that the non-sickle-cell proportion exceeds the sickle-cell proportion — that is, $d = p_2 - p_1$ exceeds zero. In addition, the user wishes to find the probability that d exceeds .5, .6, .7, .8 and .9. In the output, one types y to indicate that one wishes to compute probabilities of improvement. Then one inputs the values of interest on the DATA line. For each value in the column x, the program outputs PdAlx, the probability that the difference d is at least as large as the given value. For the value .7 in the column x, the corresponding value in the column PdAlx is the probability that $d \geq .7$. There is some small error in this probability calculation due to the simulation procedure. So in the table of probabilities, there is a sim_se column added that gives the simulation standard error for these computed probabilities. The probability that the difference d is at least as large as .7 is given by .559 plus or minus the simulation error of .016. The value of .7 is approximately the median of the posterior distribution of d since it is equally likely that the difference in proportions is smaller or larger than this value.

```
MTB > exec 'pp_beta'

FOR PROPORTION P1,
ENTER VALUES OF BETA PARAMETERS A1 AND B1:
DATA> 3 14

FOR PROPORTION P2,
ENTER VALUES OF BETA PARAMETERS A2 AND B2:
DATA> 15 2

HOW MANY VALUES OF (P1, P2) DO YOU WISH TO SIMULATE?
DATA> 1000

TYPE 'y' TO SEE A PLOT OF THE JOINT DISTRIBUTION
OF P1 AND P2:
y

TYPE 'y' TO SEE A PLOT OF THE DISTRIBUTION OF
THE DIFFERENCE IN PROPORTIONS P2-P1:
y
```

Each dot represents 4 points

```
                                         .
                                    :       :
                                    :      ..:
                                .:: ::.:::::    :.
                            :   .:::::.:::::::::::::.
                        :: :: :::::::::::::::::::::: :.
                    ....::::::::::::::::::::::::::::::.::.
        .    ...........:::..:::::::::::::::::::::::::::::::::::::.:
        -----+---------+---------+---------+---------+---------+-P2-P1
          0.36      0.48      0.60      0.72      0.84      0.96
```

TYPE 'y' to COMPUTE PROBABILITIES OF IMPROVEMENT
FOR P2-P1:

 Input values of possible improvement.
 The output is the probabilty PdALx that P2-P1 exceeds
 each improvement value x. The column sim_se gives
 simulation standard errors for the estimated probabilities.
----
y
DATA> 0 .5 .6 .7 .8 .9
DATA> end

```
ROW      x    PdALx   sim_se
  1     0.0   1.000   0.000
  2     0.5   0.951   0.007
  3     0.6   0.823   0.012
  4     0.7   0.559   0.016
  5     0.8   0.238   0.013
  6     0.9   0.024   0.005
```

5.3.3   Testing the equality of two proportions

> Minitab command to test the equality of two proportions:
> **exec 'pp_bet_t'**

As in Section 5.2.2, we consider the problem of testing the hypothesis H that two proportions p_1 and p_2 are equal against the alternative hypothesis K that the proportions are different. Both proportions are assumed to be continuous valued, so p_1 and p_2 can take on any value between 0 and 1.

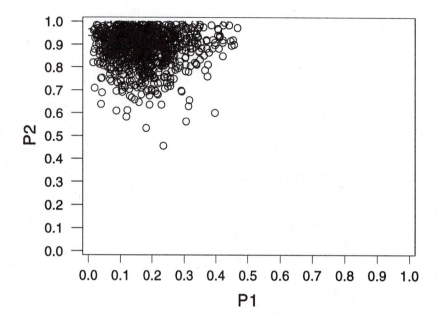

Figure 5.2: Scatterplot of simulated sample from posterior distribution of two proportions for malaria example.

As in the case of testing for a specific proportion value in Section 4.3.4, we will test the hypothesis H by finding its posterior probability. The posterior probability of H can be written as $qBF/(qBF + 1 - q)$, where q is the initial probability of H and BF is the Bayes factor, the ratio of the probabilities of the data under the hypotheses H and K. To compute this Bayesian test, three inputs will be used by the program 'pp_bet_t'. First, one specifies the prior probability of the hypothesis H. If one chooses the value .5, then this means that the hypotheses "proportions equal" and "proportions unequal" have the same initial probability of being true. Next, one has to specify two prior distributions: one if the proportions are equal (hypothesis H) and one if the proportions are different (hypothesis K). So first one thinks about possible proportion values if the proportions are equal. A beta distribution with parameters a and b will be used to model these beliefs about this unknown common proportion value. The programs described in Chapter 4 may be helpful in finding values of a and b that match one's opinion about this proportion. If little is known about its location, one can assume that the proportion is uniformly distributed (a beta density with $a = b = 1$). Last, one thinks about the locations of the two proportions if they are indeed different (hypothesis K). The program will model this opinion by means of independent beta distributions. Again, the programs of Chapter 4 can be used to help choose these beta distributions for p_1 and for p_2. If it is difficult to specify these distributions, then one can use independent uniform densities (beta(1, 1)) for the proportions.

To illustrate this Bayesian test, the command 'pp_bet_t' is run for the data in Example 1.3 from Berry[3] (see the output below). One is interested in testing if the probabilities of receiving malaria from carriers and noncarriers are equal. The following inputs reflect a "noninformative" analysis where one knows little about the location of either proportion and is unsure about the prior probabilities of the hypotheses. First one indicates that the hypothesis H of equality has a prior probability of .5. If the proportions are equal (hypothesis H), a noninformative uniform prior is chosen ($a = 1, b = 1$) for the common value of the proportions. Next, one inputs the parameters of the two beta distributions that reflect one's prior beliefs when the probabilities are not equal (hypothesis K). In this case, independent uniform priors are used; this indicates that, if $p_1 \neq p_2$, then all possible values of (p_1, p_2) in the unit square are equally likely. Last, one inputs the observed data — the number of successes and failures for the two samples. The output of this program is the Bayes factor in favor of the hypothesis H and the posterior probability that the hypothesis is true. For this example, the posterior probability of H is .00009. This value is much smaller than the prior probability, which is strong evidence that the probabilities are different.

```
MTB > exec 'pp_bet_t'

Enter the prior probability of the null hypothesis H of equality:
DATA> .5
```

---

[3]Berry, D. (1996), *Statistics: A Bayesian Perspective*, Belmont, CA.: Duxbury Press.

UNDER THE NULL HYPOTHESIS H THAT P1=P2

Enter the numbers a and b of the beta(a, b) distribution:
DATA> 1 1

UNDER THE ALTERNATIVE HYPOTHESIS K THAT P1><P2

Enter the numbers a1 and b1 of the beta(a1, b1) distribution on P1:
DATA> 1 1

Enter the numbers a2 and b2 of the beta(a2, b2) distribution on P2:
DATA> 1 1

THE DATA

Enter the number of observed successes and failures for the 1st sample:
DATA> 2 13

Enter the number of observed successes and failures for the 2nd sample:
DATA> 14 1

You have entered:
 ROW   prob_H   a_b   a1_b1   a2_b2   data
   1   0.5   1   1   1   2
   2   1   1   1   13
   3   14
   4   1

The Bayes factor in favor of the null hypothesis is:
BF_HK
  0.0000894

The Bayes factor against the null hypothesis is:
BF_KH
  11180.8

The posterior probability of the null hypothesis is:
prob_H
  0.0000894

5.3.4 Using an exchangeable prior

> Minitab command to learn about two proportions using an exchangeable prior:
> **exec 'pp_exch'**

As in the previous sections, we are again comparing two binomial proportions p_1 and p_2. In Section 5.3.2, we considered this problem when one had prior information about the location of each proportion and that the beliefs about the two proportions were independent. Beta distributions were used in that situation to represent prior opinion about p_1 and p_2.

In many situations, it can be difficult to specify prior information for each individual proportion. It can be often easier to state "structural" prior information about the relationship between the two proportions. As a simple illustration of this situation, suppose that you are interested in estimating the batting averages of two baseball players. You don't know much about baseball and have little idea about the sizes of batting averages. It would be difficult for you to specify independent beta densities for the two batting averages. However, you think that the two players have about the same batting ability and should have approximately the same batting average. In other words, your beliefs about the two proportions are *exchangeable*; your prior beliefs about the first player's batting average are about the same as your prior beliefs about the second player's batting average.

The program 'pp_exch' will construct an exchangeable prior distribution for two proportions. The belief that two proportions p_1 and p_2 are similar in size is equivalent to the belief that the corresponding logits $t_1 = \log(p_1/(1-p_1))$ and $t_2 = \log(p_2/(1-p_2))$ are similar in size. The prior belief of similarity can be modeled by means of the following hierarchical prior distribution.

1. Conditional on a mean m, the logits t_1 and t_2 are independent from the same normal distribution with mean m and standard deviation t.

2. The unknown mean m is distributed from a normal distribution with mean 0 and standard deviation 1.

Stage 1 of this prior density assigns identical normal distributions to the two logits, which reflects your belief that the two logit parameters are similar in size. However, you don't know the location of the common normal distribution. To reflect this uncertainty, the normal mean m is assigned the rather vague normal distribution with a "large" standard deviation of 1.

To use this prior distribution, one must assess one parameter — the normal standard deviation t. The value of t reflects how strongly you believe that the two proportions are similar. To see this exchangeable prior distribution for the value $t = .5$, the program 'pp_exch' is run in the output here. No data are yet observed, so we input zero successes and failures for each of the two samples. The program will simulate a large number of values from the above hierarchical prior distribution. (We type 1000 to indicate that we wish to take a simulated sample of size 1000.) The program plots the simulated values of the two proportions p_1 and p_2 using a scatterplot in Figure 5.3. Note that the points are concentrated along the diagonal line $p_1 = p_2$, which indicates that we believe that the proportions are similar in size. If we chose the smaller value $t = .1$, then the points would be more heavily concentrated about this diagonal line. The output also gives a dotplot for each proportion and for the difference in proportions $d = p_2 - p_1$. Note that the dotplots for the individual proportions are relatively flat over the entire $(0, 1)$ interval, indicating little prior knowledge about each proportion. However, the dotplot for the difference in proportions is concentrated about 0, indicating that you believe that the two proportions are about the same size.

```
MTB > exec 'pp_exch'

INPUT OBSERVED NUMBER OF SUCCESSES AND FAILURES IN 1ST GROUP:
DATA> 0 0

INPUT OBSERVED NUMBER OF SUCCESSES AND FAILURES IN 2ND GROUP:
DATA> 0 0

INPUT STANDARD DEVIATION OF THE LOGITS T1 AND T2:
DATA> .5

INPUT NUMBER OF SIMULATED VALUES:
DATA> 1000

TYPE 'y' AND RETURN TO SEE A PLOT OF THE JOINT DISTRIBUTION
OF P1 AND P2:
y
```

TYPE 'y' AND RETURN TO SEE PLOTS OF THE MARGINAL DISTRIBUTIONS
OF P1 AND P2:
y

Posterior distribution of proportion P1:
Each dot represents 3 points

```
                                    .
                         :          :   .
                       .:.  .    .    .:  :
           .    .. . ::: : ..:..:. :: ::.
           : . ::::: :::..: :::::::::.::.::::: .
         ..  :.:::::::.::::::::::::::::::::::. :
         ::.:::::::::::::::::::::::::::::::::::: .
        :.:::::::::::::::::::::::::::::::::::::::::..
      +---------+---------+---------+---------+---------+-------post_p1
     0.00      0.20      0.40      0.60      0.80      1.00
```

Variable	N	Mean	Median	TrMean	StDev	SEMean
post_p1	1000	0.50686	0.51723	0.50853	0.22380	0.00708

Variable	Min	Max	Q1	Q3
post_p1	0.03043	0.96167	0.32848	0.69874

Posterior distribution of proportion P2:
Each dot represents 3 points

```
                                 .
                       .   . :   :
              .  :    :  :.  : :   :  :
            :::..:  ::  :: .::: ..:. : :
           .:::::::..::::::.:::: ::::::.:.
        .. :::::::::::::::::::::::::::::: ..
      :..::::::::::::::::::::::::::::::::.:: .
      :..:::::::::::::::::::::::::::::::::::::.
      +---------+---------+---------+---------+---------+-------post_p2
     0.00      0.20      0.40      0.60      0.80      1.00
```

Variable	N	Mean	Median	TrMean	StDev	SEMean
post_p2	1000	0.50790	0.51143	0.50926	0.21540	0.00681

Variable	Min	Max	Q1	Q3
post_p2	0.01086	0.96513	0.33699	0.68981

TYPE 'y' AND RETURN TO SEE PLOT OF THE DISTRIBUTION OF P2-P1:
y

Each dot represents 4 points

```
                              :
                   .       .  :
                :.    ::::::.
                ::::::::::::::..
              .:.:::::::::::::::  ..
           :.::::::::::::::::::::::..:
     .  . ...:::::::::::::::::::::::::::::.:.:
  ... ...:::::.:::::::::::::::::::::::::::::::::::..:  :  .   .
   -------+---------+---------+---------+---------+---------p2-p1
       -0.32      -0.16      0.00       0.16       0.32      0.48
```

Variable	N	Mean	Median	TrMean	StDev	SEMean
p2-p1	1000	0.00104	0.00148	0.00233	0.14308	0.00452

Variable	Min	Max	Q1	Q3
p2-p1	-0.43482	0.42639	-0.09190	0.08876

What happens when we update this prior distribution with data? In our baseball example, suppose that you observe some hitting data; the first player gets four hits out of twenty at-bats and the second player gets eight out of twenty at-bats. You are interested in using your prior information about "similarity" and these data to obtain estimates of the batting averages p_1 and p_2 and to compare the averages by use of the difference $d = p_2 - p_1$.

The program is rerun to perform posterior calculations for the proportions; see the output given next. We will again use the value $t = .5$ for the standard deviation of each logit and simulate 1000 values from this prior distribution. Last, one inputs the number of successes and failures for each of the two samples. Here the first batter obtained four hits (successes) and sixteen outs (failures); the second batter had eight hits and twelve outs.

```
MTB > exec 'pp_exch'

INPUT OBSERVED NUMBER OF SUCCESSES AND FAILURES IN 1ST GROUP:
DATA> 4 16

INPUT OBSERVED NUMBER OF SUCCESSES AND FAILURES IN 2ND GROUP:
DATA> 8 12
```

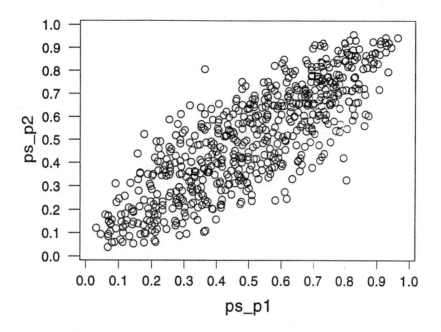

Figure 5.3: Simulated values of exchangeable prior distribution for two proportions for baseball example.

```
INPUT STANDARD DEVIATION OF THE LOGITS T1 AND T2:
DATA> .5

INPUT NUMBER OF SIMULATED VALUES:
DATA> 1000

TYPE 'y' AND RETURN TO SEE A PLOT OF THE JOINT DISTRIBUTION
OF P1 AND P2:
y

TYPE 'y' AND RETURN TO SEE PLOTS OF THE MARGINAL DISTRIBUTIONS
OF P1 AND P2:
y
```

Posterior distribution of proportion P1:

Each dot represents 6 points

```
                               :
                               :
                               :
                      :.       :         .
                     ::: :: : :  ..:.
                     :.:::.:::: :.:::.
            . .:  :::::::::::::::::::  :...
         . ..:::::::::::::::::::::::::.:::::::........
        -------+---------+---------+---------+---------+---------post_p1
           0.10      0.20      0.30      0.40      0.50      0.60
```

Variable	N	Mean	Median	TrMean	StDev	SEMean
post_p1	1000	0.26871	0.26862	0.26711	0.08350	0.00264

Variable	Min	Max	Q1	Q3
post_p1	0.06878	0.53110	0.20817	0.31934

Posterior distribution of proportion P2:
Each dot represents 5 points

```
                      :    . : :
                      :    :.: :   :
              .    :   ::: ::. :
              :    :: .::..:: ::   .
              :  :::::::::::::::.::..:
              :.:::::::::::::::::::::. :
          ... :.....:::::::::::::::::::::::::::.:.:..          .
        ---+---------+---------+---------+---------+---------+---post_p2
          0.12      0.24      0.36      0.48      0.60      0.72
```

Variable	N	Mean	Median	TrMean	StDev	SEMean
post_p2	1000	0.37429	0.36858	0.37331	0.08946	0.00283

Variable	Min	Max	Q1	Q3
post_p2	0.11251	0.70315	0.30646	0.44017

TYPE 'y' AND RETURN TO SEE PLOT OF THE DISTRIBUTION OF P2-P1:
y

Each dot represents 4 points

```
                                        :
                            :  :      ..
                          ..:  :    :::
                          :::.:: ::::.: :
                  .  :  : :::::::::::::::.:   .
                 .: :::.:::::::::::::::::::  :  : .
              .:::::.::::::::::::::::::::::: ::::  : :.
          ...  .:.....::::::::::::::::::::::::::.:::.::::.:
         -+---------+---------+---------+---------+---------+-----p2-p1
        -0.24     -0.12      0.00      0.12      0.24      0.36
```

Variable	N	Mean	Median	TrMean	StDev	SEMean
p2-p1	1000	0.10558	0.10262	0.10605	0.10904	0.00345

Variable	Min	Max	Q1	Q3
p2-p1	-0.21449	0.36028	0.03593	0.17580

The SIR simulation algorithm is used to obtain a sample from the posterior distribution of the proportions. (This algorithm will be described in more generality in Chapter 10.) First, a simulated sample of size 1000 is obtained from the prior distribution of the proportions. The simulated values are stored in the columns 'PRIOR_P1' and 'PRIOR_P2'. For each simulated value of (p_1, p_2), we compute the probability of observing the previous sample result. If the number of successes and failures in the two samples are given by (s_1, f_1) and (s_2, f_2), the likelihood is given by

$$LIKE = p_1^{s_1}(1 - p_1)^{f_1} p_2^{s_2}(1 - p_2)^{f_2}.$$

Then we take a new random sample with replacement from the prior sample of (p_1, p_2), where the sampling probabilities are proportional to the likelihoods. The resulting simulated sample comes from the posterior distribution of p_1 and p_2. The posterior sample is stored in the Minitab columns 'POST_P1' and 'POST_P2'.

This program displays a scatterplot of the simulated values of the posterior distribution for the two proportions (see Figure 5.4). The program gives descriptive statistics for the simulated values for each proportion. In addition, the program gives a dotplot and descriptive statistics for the posterior distribution of the difference in proportions $d = p_2 - p_1$. To understand this posterior distribution, recall that the observed data was four hits in twenty at-bats and eight hits in twenty at-bats. The usual estimates of

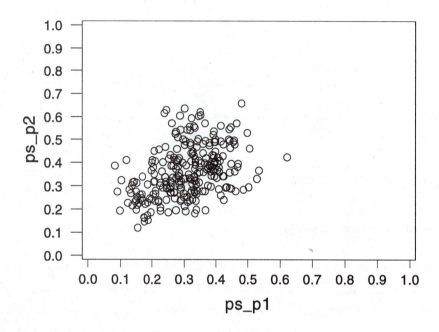

Figure 5.4: Simulated values of posterior distribution using an exchangeable prior for baseball example.

p_1 and p_2 are given by $4/20 = .2$ and $8/20 = .4$, respectively. If one thought that the hitters had identical batting averages, then one could estimate the hitting probability by pooling the data: $(4 + 8)/(20 + 20) = .3$. Here since we think that the proportions are roughly the same size, we estimate the proportions by shrinking the individual estimates .2 and .4 towards the pooled value .3. Note from the output that the medians of the simulated values from the posterior distribution of p_1 and p_2 are given by .26 and .36, respectively. By moving the individual proportion estimates towards the pooled value, we have hopefully obtained improved estimates. In addition, since we have partially pooled the two samples, one obtains more precise estimates of each batting average. From the describe commands, we see that 50% interval estimates for the two batting averages are given by (.21, .31) and (.31, .42). Last, the output is helpful in learning about the difference in batting averages $p_2 - p_1$. From looking at the dotplot and summary statistics from the simulated values of the posterior distribution of $p_2 - p_1$, we see that a good estimate at this difference in proportions is the median .11, and a 50% interval estimate is given by (.05, .17). Most of the simulated values are positive, indicating that there is good reason to think that the second player has a higher batting average.

5.4 Exercises

1. (Berry[4], exercise 8.1) One is interested in comparing the attitudes of men and women toward violence in television. The data collected from a sample of men and women are given here:

Too much TV violence?		
	Yes	No
Men	8	12
Women	17	5

Let p_M and p_W denote the population proportions who would respond yes among men and women. Assume that each proportion can take on eleven equally spaced values from 0 to 1 and assign equal probabilities to all possible pairs of values (p_M, p_W).

(a) Use the program 'pp_disc' to find the posterior distribution of the two proportions on the basis of this data.

(b) Find the probability that the difference in proportions $p_W - p_M$ exceeds each of the values 0, .3, and .6.

2. Refer to exercise 1. Suppose that you are interested in testing the hypothesis that $p_M = p_W$. Suppose as in the previous exercise that each proportion takes on eleven equally spaced values from 0 to 1. However, since you are interested in the hypothesis of equality, the prior will place total probability .5 on the pairs (p_M, p_W) where $p_M = p_W$. In addition, you will assume that each pair along the diagonal of

---

[4]Berry, D. (1996), *Statistics: A Bayesian Perspective*, Belmont, CA.: Duxbury Press.

equality will have the same probability and each pair off the diagonal will have the same probability. Using the data in Exercise 1 and the program 'pp_disct', find the posterior distribution of the two proportions and find the posterior probability that the proportions are equal.

3. (Berry[5], exercise 8.9) A study in women with node-positive breast cancer considered three chemotherapy regimes: low dose, standard dose and high dose of a combination of drugs. Among patients with a particular tumor marker, these were the results for the first three years after therapy:

	High	Standard	Low
Okay	32	25	13
Relapsed	5	16	13
Total	37	41	23

Let p_H and p_S denote the population proportions of "Okay" patients for high dose and standard dose, respectively. Suppose that the prior distribution for the two proportions is given in the following table:

			p_S			
p_H	.4	.5	.6	.7	.8	.9
.9	.01	.02	.02	.03	.03	.04
.8	.02	.02	.03	.03	.04	.03
.7	.02	.03	.03	.04	.03	.03
.6	.03	.03	.04	.03	.03	.02
.5	.03	.04	.03	.03	.02	.02
.4	.04	.03	.03	.02	.02	.01

Note that the pairs (p_H, p_S), where $p_H = p_S$ are most likely. The prior probabilities decrease as the pair moves away from the diagonal line where $p_H = p_S$.

(a) Use the program 'pp_discm' to find the posterior distribution of the two proportions.

(b) Find the posterior probabilities that $p_H < p_S$, $p_H = p_S$, and $p_H > p_S$.

[5]Berry, D. (1996), *Statistics: A Bayesian Perspective*, Belmont, CA.: Duxbury Press.

4. (Berry[6], exercise 8.12) A researcher collected aphids after spraying with two concentrations of sodium oleate. Consider models for proportions of population kill, p_L (low concentration) and for p_H (high). In assigning prior probabilities you feel that it is not possible that the low concentration has a higher kill rate than does the high, so you give zero probability to $p_L \geq p_H$. The prior probability that the proportions are equal is .22; the full table of probabilities is given here:

						p_H					
p_L	0	.1	.2	.3	.4	.5	.6	.7	.8	.9	1
1	0	0	0	0	0	0	0	0	0	0	.02
.9	0	0	0	0	0	0	0	0	0	.02	.02
.8	0	0	0	0	0	0	0	0	.02	.02	.02
.7	0	0	0	0	0	0	0	.02	.02	.02	.02
.6	0	0	0	0	0	0	.02	.02	.02	.02	.02
.5	0	0	0	0	0	.02	.02	.02	.02	.02	.01
.4	0	0	0	0	.02	.02	.02	.02	.01	.01	.01
.3	0	0	0	.02	.02	.02	.02	.01	.01	.01	.01
.2	0	0	.02	.02	.02	.01	.01	.01	.01	.01	.01
.1	0	.02	.02	.01	.01	.01	.01	.01	.01	.01	.01
0	.02	.01	.01	.01	.01	.01	.01	.01	.01	.01	.01

Suppose that 68 aphids were sprayed with the low concentration spray and 55 died and 13 were alive; of 65 sprayed with the high concentration spray 62 died and 3 were alive.

(a) Find the posterior probabilities of all pairs of models for (p_L, p_H).

(b) Find the posterior probability that the difference in proportions $p_H - p_L$ exceeds each of the values 0, .1, .2, .3, .4.

5. (Berry[6], exercise 9.9) A randomized study in 59 centers in the United States addressed the effectiveness of the drug AZT in preventing transmission of AIDS-associated virus HIV to newborns from infected mothers. Of newborns in the AZT group, 13 of 157 tested positive for HIV. Of newborns in the placebo group, 40 of 157 tested positive. Consider the proportions of those who test positive

[6]Berry, D. (1996), *Statistics: A Bayesian Perspective*, Belmont, CA.: Duxbury Press.

in the corresponding populations: AZT (T) and placebo (C), and the difference $d = p_T - p_C$. Assuming a beta(1, 1) prior for both p_T and p_C:

(a) Find a 95% posterior probability interval for d.

(b) Find the probability that $d < 0$ (the treatment is beneficial).

(c) Recompute (a) and (b) in the case where a beta(2, 18) distribution is assigned to p_T and a beta(25, 75) distribution is assigned to p_C. (These priors reflect the belief that AZT has a beneficial effect.)

6. (Berry[7], exercise 9.11) Between 1980 and 1989, fifty-eight abstracts dealing with maternal cocaine use and fetal health problems were submitted for presentation at annual meetings of the Society for Pediatric Research. Nine of the 58 (the negative reports) reported no adverse effects·on babies of cocaine users, and the other 49 (the positive reports) described reproductive risks linked to cocaine use. Peer reviewers accepted only one of the first nine for presentation while they accepted 28 of the latter 49. Is this sufficient information to infer that an abstract's conclusion affects its acceptability to peers in this profession? Let p_N denote the probability a negative report is accepted and p_P the probability a positive report is accepted. Assuming beta(1, 1) priors for these two proportions:

(a) Find a 90% posterior probability interval for the difference $d = p_P - p_N$.

(b) Using the program 'pp_beta_t', test the hypothesis that $p_P = p_N$. Assign the hypothesis of equality a prior probability of .5. Assign the common value of p a beta(1, 1) prior under the hypothesis of equality and assign the two proportions beta(1, 1) priors under the alternative hypothesis that the proportions are unequal.

7. (Berry[7], exercise 9.15) A survey was taken to assess whether freshmen college underweight women were more likely to subscribe to fashion or beauty magazines. Of the 36 underweight women, 16 subscribed to one of six fashion or beauty magazines; of 39 women who were not underweight, 14 subscribed to one of these magazines. Regard these to be random samples from the populations of

[7]Berry, D. (1996), *Statistics: A Bayesian Perspective*, Belmont, CA.: Duxbury Press.

underweight and not underweight freshmen women. Let p_U and p_N denote the proportions from these two groups that subscribe to the magazines, and suppose that a beta(1, 2) distribution models one opinion about each proportion.

(a) If $d = p_U - p_N$ denotes the difference in proportions, find the posterior probability that the d is at least as large as 0, .1, and .2.

(b) Find a 90% posterior probability interval for d.

(c) Test the hypothesis that the proportions are equal. Run the program 'pp_beta_t' twice, using different sets of beta prior distributions for the proportions under the two hypotheses.

8. (Berry[8], exercise 9.22) Three students were interested in whether basketball players are more effective when under less pressure. They considered the three-point shots attempted by Duke's 1992–1993 basketball team in the first half versus second half of games, thinking that there would be more pressure in the second half. Let p_F (p_S) denote the population proportions of three-point shots made in the first (second) half. Of the 211 three-point shots attempted in the first half, 71 were successful; of the 255 attempted in the second half, 90 were successful.

(a) Assuming beta(1, 1) priors for each proportion, find a 90% posterior probability interval for the difference $d = p_S - p_F$.

(b) Test the hypothesis that the two proportions are equal using the program 'pp_bet_t'. Run this program twice. The first time use uniform (beta(1, 1)) prior distributions for the proportions under both hypotheses and the second time use prior distributions that better reflect your knowledge about the locations of the proportions.

9. (Berry[8], exercise 9.33) A company is concerned about a defect in one of its products. The defect cannot be found on the assembly line but must be tested in real or simulated conditions. A total of fifty-six items from two different manufacturing lots were tested. In forty items from lot 1, eleven were found defective, and one out of sixteen items from lot 2 were defective.

---

[8]Berry, D. (1996), *Statistics: A Bayesian Perspective*, Belmont, CA.: Duxbury Press.

(a) Suppose that you believe that the proportions of defectives from the two manufacturing lots are exchangeable. That is, your opinion about the chances of defectives in the first lot is the same as your opinion about the likelihood of defectives in the second lot. Use the program 'pp_exch' to estimate the difference in defective proportions $d = p_2 - p_1$, where $p_1 (p_2)$ is the proportion of defectives in lot 1 (lot 2). (In this run, let the standard deviation of the logits be equal to .5.) To obtain a simulated sample of the difference of proportions, use the Minitab commands

```
name c3 'd'
let 'd'='post_p2'-'post_p1'
```

(b) Is there sufficient evidence to conclude that the defective proportions are different? Construct a Bayes test of the hypothesis $p_1 = p_2$ using the program 'pp_bet_t'. Assume uniform priors under both the hypothesis of equality and the hypothesis that the proportions are different.

Chapter 6

Learning about a Normal Mean

6.1   Introduction

This chapter describes a number of Minitab programs to learn about a population mean. Suppose that each member of the population of interest has an associated continuous measurement. We assume that the histogram of measurements for the entire population is normally distributed with mean M and standard deviation S. The primary goal of the programs in this chapter is to learn about the population mean M.

To illustrate this situation, consider the following example from Chapter 11 of Berry[1]. Berry was interested in determining his true weight from a bathroom scale that gives readings that appear to be quite variable. Here the population would be the collection of measurements from the scale if Berry were able to weigh himself an infinite number of times. We assume that these measurements are normally distributed with mean M and standard deviation S. If the scale is not biased in either direction, then the population M would correspond to Berry's true weight.

To learn about his true weight, Berry weighs himself ten times, obtaining the measurements 182, 172, 173, 176, 176, 180, 173, 174, 179, 175. After he weighs himself, a number of questions come to mind. What is his true weight? Berry can learn about his true weight by means of an "interval estimate" that contains M with a high probability. Suppose that his doctor says that Berry is overweight if he weighs more than 180 pounds. Is he overweight? To answer this question, Berry will want to compute

---

[1]Berry, D. (1996), *Statistics: A Bayesian Perspective*, Belmont, CA.: Duxbury Press.

the probability that his true weight is larger than 180. If this probability is sufficiently small, he can conclude that he is not overweight. What if Berry takes a new measurement tomorrow from the scale? Can he make a reasonable prediction about the value of this new measurement? Berry may wish to construct a prediction interval that he believes will contain this new measurement with a given probability.

The Minitab programs in this chapter are useful for answering the earlier questions. They can be used for learning about a population mean, testing or making a decision about its value, and predicting the values of a future measurement. The programs can be categorized on how one represents prior opinion about the unknown population mean and standard deviation. The simplest approach, implemented by the program 'm_disc', assumes that the population standard deviation S is known and constructs a prior distribution for M using the discrete approach that was used for proportions in Chapters 4 and 5. One specifies a set of plausible models for M, assigns probabilities to these models, and updates these probabilities using Bayes' rule.

The remainder of the programs use a continuous approach for modeling beliefs about the population mean. One can assume that Berry's true weight M is a continuous unknown quantity. A convenient method of modeling information about Berry's weight is by use of a normal density curve. The program 'normal_s' can be used to find the particular normal density for M that matches statements by the user about two percentiles of the prior density. After this normal prior density has been found, the next step is to update one's opinion about M from the ten measurements. The program 'm_cont' is useful for this updating. One inputs the prior density and the data, and the program gives the mean and standard deviation of the approximate normal density for the population mean M. In addition, this program gives an approximate prediction density for a future measurement.

One advantage of the approximate analysis of 'm_cont' is that it makes a convenient correction for the fact that the standard deviation for the population S is unknown. Since the values for both the mean M and the standard deviation S are uncertain, a more standard analysis would assign a joint prior density to these two parameters and estimate M from its marginal posterior density. Since this joint prior density is difficult to assess, the program 'm_nchi' finds the marginal posterior densities for both M and S in the case when an noninformative prior density is used. These marginal posterior densities are of standard functional forms, so this program implements a more exact analysis that the one used in 'm_cont'. This program could be used for our example if Berry had little prior knowledge about his true weight.

The programs 'm_cont' and 'm_nchi' focus on the estimation and prediction problems. To test the hypothesis that the mean M is equal to a particular value of interest, the program 'm_norm_t' is used. One is interested in testing the hypothesis H that M is equal to a particular value, say 180, against the alternative hypothesis that M is not equal to this value. To construct this test, one must first specify the prior probability that H is true. In addition, if M is not equal to 180, one specifies a normal density that indicates possible alternative values for the mean. The output of this program is the posterior probability of the hypothesis H and the Bayes factor in support of this hypothesis.

6.2 Using discrete models

> Minitab command to learn about a mean using discrete models:
> **exec 'm_disc'**

Suppose in our example that Berry doesn't know his exact weight, but he thinks it is either 174, 176, or 178 pounds. Moreover, he believes before seeing any measurements that each of the three values is equally likely. We make the assumption that the population of measurements of Berry's scale is normally distributed with mean M that is equal to his actual weight. Berry does know (from previous experience with the scale) that the standard deviation of the population of measurements is approximately three pounds.

The command 'm_disc' implements Bayes' rule for a collection of normal mean models where the standard deviation of the population is known (see the output here). Before this program is run, two Minitab columns named 'm' and 'prior' are defined. The values of the population mean M are placed in the column 'm' and the respective prior probabilities are placed in the column 'prior'. Last, one puts the ten observed measurements in the worksheet. In this output, the ten weighings are placed in column C3, which is named 'data'.

```
MTB > name c1 'm' c2 'prior'
MTB > set 'm'
DATA> 174 176 178
DATA> end
MTB > set 'prior'
DATA> .333 .333 .333
DATA> end
MTB > name c3 'data'
MTB > set 'data'
DATA> 182 172 173 176 176 180 173 174 179 175
DATA> end
MTB > exec 'm_disc'

INPUT POPULATION STANDARD DEVIATION:
DATA> 3

OBSERVED DATA IN WORKSHEET? (TYPE 'y' OR 'n'.)
  IF YES, INPUT NUMBER OF COLUMN.
  IF NO, INPUT OBSERVED SAMPLE MEAN AND SAMPLE SIZE:
y
DATA> 3

OBS_DATA
    182     172     173     176     176     180     173     174     179     175
```

Row	m	prior	M_x_PRIO	LIKE	PRODUCT	POST	M_x_POST
1	174	0.333333	58.000	108368	36123	0.089065	15.497
2	176	0.333333	58.667	1000000	333333	0.821871	144.649
3	178	0.333333	59.333	108368	36123	0.089065	15.853
4			176.000				176.000

After the 'm' and 'prior' and data columns have been defined, the program 'm_disc' can be run. There are two inputs to this program — the value of the known standard deviation S and the data. The data can be entered in one of two ways. They can be entered in "summary form" (sample mean and sample size), or they can be located in a column of the worksheet. Here we indicate that the data are in the worksheet and type 3 to indicate that the data is located in column C3.

The program implements Bayes' rule in the familiar table format that was used for the proportion problem in Chapter 4. The values of the posterior probabilities are stored in the column 'POST'. Note that, after observing these ten measurements, the posterior

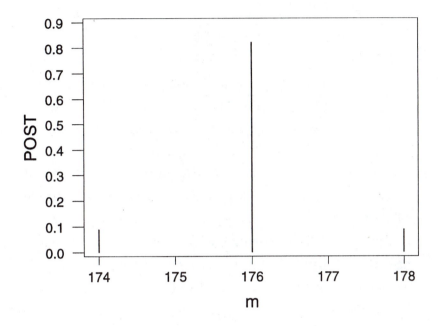

Figure 6.1: Posterior probabilities of mean M for weighing example.

probability that your actual weight is 176 is over 80%, although the two alternative values of 174 and 178 are still plausible. The line graph displayed by this program (see Figure 6.1) dramatically shows the concentration of the posterior probability in this case. This program also gives the prior and posterior means for M. In this example, both prior and posterior distributions have a mean of 176. The prior and posterior distributions differ in that the posterior distribution is more concentrated about its mean.

6.3   Using continuous models

6.3.1   Assessing a prior density

Minitab command to assess a normal prior density:
exec 'normal_s'

Let's reconsider the weighing example, but with the use of continuous models for the population mean. Suppose that Berry believes that his actual weight M can take on a continuous range of values. He would like to find a prior density for M that roughly reflects his opinion about his current weight. The program 'normal_s' can be used to find a normal density that matches the specification of two percentiles of the distribution. A percentile is a particular value for M such that a given probability is smaller than that value. Berry thinks that his true weight is equally likely to fall above or below 174. Equivalently, the .5 percentile of the distribution of M is equal to 174. In addition, he thinks that the chance that his weight is larger than 180 is unlikely — only about 10%. That is, the .9 percentile of the distribution is equal to 180.

To find a normal prior density that matches this information about two percentiles, one runs the program 'normal_s'. The run of this command for this example is displayed here. One inputs a percentile of the prior distribution by specifying two numbers: q, the probability value to the left, and m, the value of M such that $P(M$ is less than $m)$ is equal to q. The program gives the mean and standard deviation of the normal distribution that matches these two percentiles. Since there may be some error in Berry's specification of these two percentiles, a different normal distribution may better reflect his prior beliefs about M. So we recommend that Berry runs the program 'normal' (to be described soon) to learn more about his choice of prior distribution and perhaps make adjustments to the prior mean and standard deviation.

```
MTB > exec 'normal_s'

Input the first probability P1 and percentile M1:
DATA> .5 174

Input the second probability P2 and percentile M2:
DATA> .9 180
```

```
The matching values of the normal density parameters
corresponding to your 2 percentiles are given by:
MEAN
    174
STD
   4.68183
```

6.3.2 Computing the posterior density

> Minitab command to compute the posterior density for a mean:
> **exec 'm_cont'**

The program 'm_cont' is used to make posterior inferences about a normal mean and predictive inference about a future observation when a normal prior is used to model beliefs about M. In the weighing example, Berry takes ten measurements of his weight from the scale. He would like to update his normal prior density for his true weight M with this new information. In addition, he plans to weigh himself one more time. He would like to predict the reading on the scale for this new weighing.

Suppose that these ten measurements have been placed in column C3 of the worksheet. The program 'm_cont' is run, and the output from this run is shown here. The program first asks if a flat prior density is to be used. If the user has little information about the location of the mean M, then one would use a prior density which has a flat shape over a wide range of M values. This is not the case here, since Berry had some opinions about his true weight. So we answer "n" and then input the mean and standard deviation for the normal prior density. The program next asks about the data. There are two ways to enter data into the program: in summary form (sample mean, sample standard deviation, and sample size) or in a particular column of the worksheet. It is indicated below that the data is in the worksheet and one inputs "3" for the number of the column that contains the data.

```
MTB > exec 'm_cont'

DO YOU WISH TO USE A FLAT PRIOR DENSITY FOR M? (TYPE 'y' OR 'n'.)
   IF NO, INPUT MEAN AND STANDARD DEVIATION FOR THE PRIOR DENSITY..
n
DATA> 174 4.68
```

```
PR_MEAN
   174

PR_STD
  4.68

OBSERVED DATA IN WORKSHEET? (TYPE 'y' OR 'n'.)
  IF YES, INPUT NUMBER OF COLUMN.
  IF NO, INPUT OBSERVED SAMPLE MEAN, STANDARD DEVIATION, AND SAMPLE SIZE.
y
DATA> 3

OBS_DATA
   182    172    173    176    176    180    173    174    179    175

THE POSTERIOR DENSITY FOR M IS NORMAL
WITH MEAN AND STANDARD DEVIATION:

MEAN STD
  175.877     1.162

THE PREDICTIVE DENSITY OF THE NEXT OBSERVATION
IS NORMAL WITH MEAN AND STANDARD DEVIATION:

MEAN STD
  175.877     3.969
```

The output of the program are the means and standard deviations for the normal posterior density for the mean M and for the normal predictive density for a future observation. These two normal densities are actually approximations for the exact posterior and predictive distributions. However, the approximations are generally very accurate. One nice feature of these approximations is that the normal density forms are easy to summarize. In this example, the updated distribution for Berry's actual weight M is normal with mean 175.88 and standard deviation 1.16. In addition, if Berry takes one additional reading from the scale, his opinion about this unknown future measurement is reflected by a normal distribution with mean 175.88 and standard deviation 3.97.

6.3.3 Summarizing the posterior density

> Minitab command to summarize a normal density:
> **exec 'normal'**

The program 'm_cont' is used for finding the posterior and predictive distributions of interest. To summarize these normal distributions, the program 'normal' can be used. This program is similar in style to the program 'p_beta' for summarizing a beta distribution. The output here displays a run of this program for learning more about the normal posterior distribution for M. In the run, one inputs the mean and standard deviation of the normal density. The program provides three types of summaries. First, if desired, the program will graph the normal density curve. It is easy to see from the normal curve in Figure 6.2 that one is pretty certain that Berry's true weight lies between 173 and 179. The program will also compute any cumulative probabilities or percentiles of interest. Suppose that Berry is interested in computing the probability that M is smaller than the values 170, 180, and 190 and finding a 95% probability interval for M. One types y to indicate that one is interested in computing cumulative probabilities and then places the values 170, 180 and 190 on the DATA line (an end command is entered on the following line to indicate that there are no further values to be entered). The program then gives the corresponding cumulative probabilities. In a similar fashion, one can find a 95% probability interval by finding the .025 and .975 percentiles of the normal density. From the output of this program, we find that $P(M < 170) = 0$ and $P(M < 180) = .99981$. Thus Berry is almost certain that his actual weight is between 170 and 180. Moreover, by the computation of the percentiles, we see that the probability is .95 that M falls in the interval (173.6, 178.2).

```
MTB > exec 'normal'

INPUT THE VALUES OF THE MEAN AND STANDARD
DEVIATION OF THE NORMAL DISTRIBUTION:
DATA>    175.877    1.162

Your values of the mean and standard deviation are:

Row    MEAN    STD
  1   175.877  1.162
```

```
TYPE 'y' TO SEE A PLOT OF THE NORMAL DENSITY:
y

TYPE 'y' TO COMPUTE CUMULATIVE PROBABILITIES:
-----------------------------------------------------------------
  Input values of M of interest.  The output is the column of
  values M and the column of cumulative probabilities PROB_LT.
-----------------------------------------------------------------
y
DATA> 170 180 190
DATA> end

  Row      M    PROB_LT
    1     170   0.00000
    2     180   0.99981
    3     190   1.00000

TYPE 'y' TO COMPUTE QUANTILES:
-----------------------------------------------------------------
  Input probabilities for which you wish to compute quantiles.
  The output is the probabilities in the column PROB and the
  corresponding quantiles in the column QUANTILE.
-----------------------------------------------------------------
y
DATA> .025 .975
DATA> end

  Row    PROB   QUANTILE
    1   0.025   173.600
    2   0.975   178.154
```

6.4   A test for a normal mean

> Minitab command to test for a normal mean:
> **exec 'm_norm_t'**

The program 'm_norm_t' performs a Bayesian test of the hypothesis that a mean from a normal distribution (with known standard deviation) is equal to a specific value. In the weighing example, suppose that Berry knows that his weight last year was 170 pounds and he wonders whether he still weighs 170 this year. So he is interested in the hypothesis H that his true current weight M is equal to 170. The alternative hypothesis K is that his weight is now either larger or smaller than 170. As in the previous illustrations

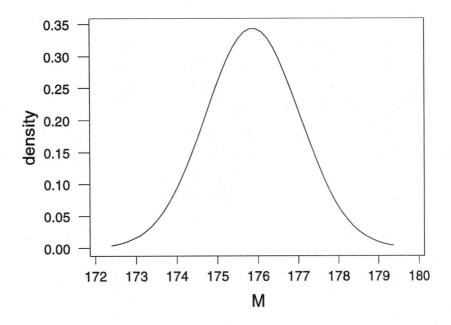

Figure 6.2: Normal (175.88, 1.16) posterior density for weighing example.

of Bayesian tests, one will decide whether the hypothesis H is true by computing its posterior probability. One useful statistic is the Bayes factor that is the ratio of the probability of the observed data under the hypothesis H to the probability of the data under K. The run of the program 'm_norm_t' for the weighing data is shown here. First one inputs the value of the population mean M to be tested and the prior probability that this hypothesis is true. Here Berry is testing the hypothesis $M = 170$. He believes that there is a good chance that his weight did not change from last year, and so he assigns this statement a probability of .5. Next, one has to think about plausible values for M if the hypothesis H is not true. If his weight did change from last year, then Berry thinks that it is more likely that M is close to last year's weight (170) than far from it. A normal distribution with mean 170 and standard deviation h will then be a suitable model for alternative values for M. In general, a normal prior distribution with mean M_0 and standard deviation h will be used to represent one's opinion under the alternative hypothesis K, where M_0 is the value to be tested.

To complete the specification of the prior distribution, one has to input the standard deviation h of the normal density under the alternative hypothesis K. If Berry's weight did change from last year, how large will be the change? One way of obtaining the value of h is to think of the range of possible alternative values for M, and then solving for this standard deviation by setting the 95% range of the normal distribution, 4 h, to this range. To illustrate, suppose that Berry thinks that his current weight could be five pounds less or more than last year's weight of 170. The range of alternative values for M is $175 - 165 = 10$ and by setting $10 = 4\,h$, one obtains $h = 2.5$.

```
MTB > exec 'm_norm_t'

ENTER THE NULL HYPOTHESIS MEAN MO:
DATA> 170

ENTER THE PRIOR PROBABILITY OF MO:
DATA> .5

FOR THE ALTERNATIVE HYPOTHESIS THAT M = MO,
ENTER STANDARD DEVIATION(S) OF THE NORMAL PRIOR DISTRIBUTION:
DATA> .5 1 2 4 8
DATA> end

ENTER THE STANDARD DEVIATION OF THE POPULATION:
DATA> 3

OBSERVED DATA IN WORKSHEET? (TYPE 'y' OR 'n'.)
  IF YES, INPUT NUMBER OF COLUMN.
  IF NO, INPUT OBSERVED SAMPLE MEAN AND SAMPLE SIZE:
y
DATA> 3

OBS_DATA
    182     172     173     176     176     180     173     174     179     175

  Mean of OBS_DATA=        176.00
  Total number of observations in OBS_DATA=        10

The Bayes factor in favor of the null hypothesis is:
BF_HK
  0.000000   0.000092   0.018964   0.138641   0.155533
```

```
The Bayes factor against the null hypothesis is:
BF_KH
   4070108      10848        53         7         6
```

```
The posterior probability of the null hypothesis is:
prob_H
   0.000000    0.000092    0.018611   0.121760   0.134598
```

Since it may be difficult to assess values for h, the program will allow the user to input a set of plausible values. In the output, the values .5, 1, 2, 4, 8 are inputted as possible values for h. Last, the program asks for the value for the known standard deviation and the data.

For each value of the prior standard deviation h, the program gives the Bayes factor in support of the hypothesis that M takes on the specific value, the Bayes factor in support of the alternative hypothesis, and the posterior probability that the hypothesis H is true. If Berry uses a normal (170, 2) density to reflect alternative values for his weight M, then the Bayes factor in support of the hypothesis $M = 170$ is equal to .018964. The posterior probability that his weight hasn't changed is .018611, which is much smaller than Berry's prior probability of .5. He should conclude that his current weight is not 170.

6.5 Exact inference about a continuous normal mean

> Minitab command to perform exact inference about a normal mean:
> **exec 'm_nchi'**

The command 'm_cont' described in Section 6.3 implements an analysis that assumes that the posterior density for M is suitably approximated by a normal curve. The command 'm_nchi' implements an exact analysis for the mean M and standard deviation S of a normal distribution. It is relatively difficult to assign an informative joint prior distribution for two parameters. So we assume that little prior information exists about M and S and so they are assigned the standard noninformative prior distribution proportional to $1/S$. In this case, the mean M has a t marginal posterior distribution and S has a marginal distribution that is closely related to a chi-squared form.

The run of this program is shown for the weighing data. The inputs are y (the data is in the worksheet) and the number of the data column. The output is graphs of the marginal posterior density for M and S (Figures 6.3 and 6.4) together with some summary information about the marginal posterior distributions. The program gives the location, scale, and degrees of freedom for the marginal t distribution for M and the scale factor and degrees of freedom for the scaled-chi distribution for the standard deviation S. Last, intervals that contain 95% of the probability for M and S are given.

Since two methods have been presented for learning about a continuous-valued mean M, it is instructive to compare the interval limits for M using the two programs. Using the approximate normal analysis and an informative prior density, the 95% probability interval for Berry's true weight was (173.5, 178.3). The exact 95% probability interval based on a t marginal posterior density and a noninformative prior is (173.6, 178.4). The approximate interval is a little shorter than the exact interval, which can be the effect of the extra information provided by the prior density. But it appears that Berry's prior information had relatively little impact on the final inference about M. Also, this brief analysis suggests that the programs 'm_cont' and 'm_nchi' will provide similar posterior inferences in the case where the information in the data is significantly larger than the information in the prior distribution.

```
MTB > exec 'm_nchi'

OBSERVED DATA IN WORKSHEET? (TYPE 'y' OR 'n'.)
  IF YES, INPUT NUMBER OF COLUMN.
  IF NO, INPUT OBSERVED SAMPLE MEAN, STANDARD DEVIATION, AND SAMPLE SIZE.
y
DATA> 3

The mean M has a t(m,se,df) distribution with

  Row     m        se    df
   1     176    1.05409   9

A 95% probability interval for M is:

m_int
   173.615    178.385
```

A 95% probability interval for S is:

```
s_int
  2.29278    6.08537
```

Type 'y' and return to see plots of marginal posterior densities for M and
 S:
y

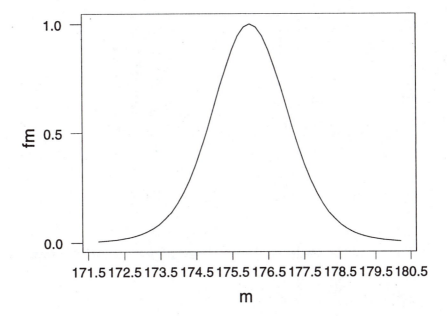

Figure 6.3: Posterior density of mean M for weighing example.

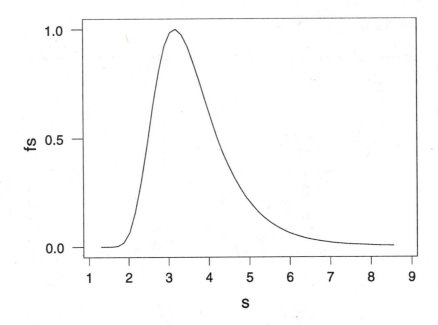

Figure 6.4: Posterior density of standard deviation S for weighing example.

6.6   Exercises

1. (Berry[2], exercise 10.8) A method for weighing extremely light objects gives the following results (in micrograms, μg) for nine weighings of a particular specimen:

$$114, 129, 121, 98, 140, 134, 122, 133, 125$$

Suppose that the specimen actually weighs M μg and that the population from which this sample is drawn is normal(M, 10). Assume prior probabilities of 1/3 for each of $M = 110$, 120, and 130. Calculate the posterior probabilities for these three values of M.

---

[2]Berry, D. (1996), *Statistics: A Bayesian Perspective*, Belmont, CA.: Duxbury Press.

2. (Berry[3], exercise 10.9) Darwin once carried out an experiment to learn whether cross- or self-pollination would produce superior plants. The following measurements are the difference in heights (in eighths of inches) between the cross- and self-pollinated plants in each pair:

$$49, -67, 8, 16, 6, 23, 28, 41, 14, 29, 56, 24, 75, 60, -48$$

(a) Suppose the model that produced these differences is a member of a set of forty normal densities, all having standard deviation 30 and with one normal density for each mean $M = -100, -95, -90, ..., 95$. Assign each of these models prior probability 1/40. Calculate the posterior probabilities for these forty models.

(b) Find a set of M values that contains approximately 95% of the posterior probability.

3. (Berry[3], exercise 11.22) In an experiment to test observational learning, sixteen octopuses watched other octopuses that had been trained to select red balls when offered a choice between red and white. Each octopus was then given five opportunities in the same setting. The average number of reds selected was 4.31 with a standard deviation .98; the corresponding average proportion of reds selected was .862 with a standard deviation of .196. Assume that these sixteen measurements are a random sample from the population of octopuses allowed to make similar observations and selections and that the mean proportion of correct selections is M.

(a) Assuming a noninformative prior, find a 95% posterior probability interval for M. Use both the approximate and exact methods (programs 'm_cont' and 'm_nchi') and compare the results.

(b) Suppose that from prior experience, you can assign M a normal distribution with mean .8 and standard deviation .1. Find a 95% posterior probability interval for M.

---

[3] Berry, D. (1996), *Statistics: A Bayesian Perspective*, Belmont, CA.: Duxbury Press.

4. Refer back to exercise 2. Suppose that you assume M is continuous with a non-informative prior. Find a 95% posterior probability interval for M and compare your result with the discrete model analysis that you applied in exercise 2.

5. Suppose a student takes an IQ test five times with the scores 110, 105, 110, 120, 105. If M is the student's "true" IQ, then one can assume that the test scores are normally distributed with mean M. Suppose that the standard deviation of this distribution of measurements is 10. Using the program 'm_cont_t', test the hypothesis that the student has an average IQ of $M = 100$. This program requires you to give the prior standard deviation of M if the student is not average. Try different values for this standard deviation, and compute the corresponding Bayes factors and posterior probabilities that $M = 100$.

6. (O'Hagen[4], exercise 9.c.1) A physicist is interested in estimating $M = $ average length of a polymer (in millimeters). He measures his prior information as normal with mean 100 and standard deviation 10. In an experiment, he isolates five polymer strands, straightens and measures them. The measurements are 77, 65, 91, 85, 78, which he assumes are normally distributed with mean M. Obtain his posterior distribution for M. Can he now claim that M is smaller than 100?

7. Suppose you are interested in learning about the mean pulse rate M (in beats per minute) of college students. Assume that the distribution of pulse rates for all students is approximately normally distributed with mean M and standard deviation 10. Suppose that the possible values for M are all integer values between sixty and ninety. Assign the same prior probability to each value.

   (a) Measure your own pulse (number of beats for one minute) and ask a few of your friends to do the same. Place the pulse measurements in a column of the worksheet.

   (b) Run the program 'm_disc' to learn about M. What value of M has the largest probability? What is the probability that the mean pulse rate is larger than seventy? Give a list of values for M that contains most of the posterior

---

[4]O'Hagan, A. (1988), *Probability: Methods and Measurement*, New York: Chapman and Hall.

probabilities. (You can include the values of M where the probability is larger than .001.)

8. Consider the problem of estimating the average total snowfall per year M (in inches) for a large city on the East Coast of the United States. Suppose that the individual yearly snow totals are approximately normally distributed about the average M.

    (a) Use the program 'normal_s' to construct a prior density for M. Guess at the median (the .5 percentile) and the .9 percentile for this average total snowfall.

    (b) The total snowfalls of this particular city for twelve years between 1962 and 1973 are given here. Place this data into the Minitab worksheet and use the program 'm_cont' to update your beliefs about the average snowfall.

    $$38.6, 42.4, 57.5, 40.5, 51.7, 67.1, 33.4, 60.9, 64.1, 40.1, 40.7, 6.4$$

    (c) Find a 95% probability interval for the average snowfall M.

    (d) What is the probability that next year's total snowfall will exceed 40 inches?

9. You are interested in estimating the average number of brown candies M in the "fun-size" bag of M&M's. Each bag contains approximately twenty-four candies, and the possible colors of the candies are red, yellow, blue, brown, and green.

    (a) Construct a prior density for M using the program 'normal_s'. Guess at the prior median and 90th percentile of this average number of brown candies.

    (b) Students in a statistics class counted the number of brown candies in 10 bags of M&M's. The counts are given here. Put this data into the Minitab worksheet.

    $$8, 9, 5, 2, 4, 6, 8, 7, 10, 8$$

    (c) Update your normal prior density using the program 'm_cont'. Find a 90% probability interval for the average number M.

    (d) If you open a new bag of M&M's, predict how many browns you will find. Find an interval that you are 90% confident will contain the number of browns.

Chapter 7

Learning about Two Normal Means

7.1 Introduction

The Minitab programs in this chapter are used to compare two normal means from data collected from two independent samples. Suppose that there are two populations of interest, and associated with each member of the population is a continuous measurement. Assume that the group of measurements from each population is normally distributed. The objective of this chapter is to compare the mean measurement M_1 of the first population with the mean measurement M_2 from the second population. To learn about the means, random samples are taken from each population. We assume that the samples are independent; informally, this means that the random process of choosing the second sample is unrelated to the choice of the first sample.

To illustrate these commands, we consider an example from Berry[1]. Birthweights of children were collected from mothers who smoke cigarettes and mothers who once smoked but quit. There are two hypothetical populations, the birthweights of all mothers who smoke cigarettes and the birthweights of all mothers who used to smoke. It is assumed that the histograms of birthweights from the "smoking" and "quit" populations are approximately normally distributed with respective means M_1 and M_2. To learn about the benefit of quitting, we are interested in the difference in population means $d = M_2 - M_1$, which represents the average benefit of quitting cigarettes.

---

[1]Berry, D. (1996), *Statistics: A Bayesian Perspective*, Belmont, CA.: Duxbury Press.

In Berry's example, prior information was available about the sizes of each population mean. The prior beliefs for both M_1 and M_2 were each modeled using a normal distribution with mean 7.7 and standard deviation .7. In addition, one's beliefs about one population mean are independent of any knowledge about the second mean. This prior information reflects the knowledge that most birthweights are in the six to nine pound range. In addition, since the prior densities for both means are identical, this choice indicates that one believes that quitting does not have an effect on the average birthweight.

7.2   An analysis based on normal distributions

> Minitab command to learn about two means using continuous models
> (approximate method):
> **exec 'mm_cont'**

The Minitab program 'mm_cont' is used together with the program 'm_cont' to learn about the difference in population means when data are collected from two independent samples. The program 'm_cont' will approximate the posterior density of the first population mean M_1 by a normal distribution. Likewise, the posterior density of M_2 will be approximately normally distributed. If one's prior beliefs about the two means are independent and data are collected from two independent samples, then the posterior distributions of the means will be independent. Then, due to this independence assumption, the posterior density of the difference in means $d = M_2 - M_1$ will also be normally distributed.

In the following output, the birthweights from the mothers who smoked is placed in column C1 and the birthweights from the mothers who quit is placed in column C2. The program 'm_cont' is run twice to learn about each of the two population means. In the run of 'm_cont', the number of the data column and the mean and standard deviation of the normal prior density are entered. Looking at the output, we see that the mean birthweight from smoking mothers is approximately normal(6.92, .23), and the mean birthweight from mothers who quit smoking is normal(6.92, .25).

```
MTB > name c1 'smoke' c2 'quit'
MTB > set 'smoke'
DATA> 4.5 5.4 5.6 5.9 6 6.1 6.4 6.6 6.6 6.6 6.9 6.9
```

```
DATA> 7.1 7.1 7.2 7.5 7.6 7.6 7.8 8 9.9
DATA> end
MTB > set 'quit'
DATA> 5.4 6.6 6.8 6.8 6.9 7.2 7.3 7.4
DATA> end
MTB > exec 'm_cont'
```

```
DO YOU WISH TO USE A FLAT PRIOR DENSITY FOR M? (TYPE 'y' OR 'n'.)
   IF NO, INPUT MEAN AND STANDARD DEVIATION FOR THE PRIOR DENSITY..
n
DATA> 7.7 .7
```

```
OBSERVED DATA IN WORKSHEET? (TYPE 'y' OR 'n'.)
   IF YES, INPUT NUMBER OF COLUMN.
   IF NO, INPUT OBSERVED SAMPLE MEAN, STANDARD DEVIATION, AND SAMPLE SIZE.
y
DATA> 1
```

```
THE POSTERIOR DENSITY FOR M IS NORMAL
WITH MEAN AND STANDARD DEVIATION:
MEAN STD
   6.92242   0.23484
```

```
THE PREDICTIVE DENSITY OF THE NEXT OBSERVATION
IS NORMAL WITH MEAN AND STANDARD DEVIATION:
MEAN STD
   6.92242   1.16624
```

```
MTB > exec 'm_cont'
```

```
DO YOU WISH TO USE A FLAT PRIOR DENSITY FOR M? (TYPE 'y' OR 'n'.)
   IF NO, INPUT MEAN AND STANDARD DEVIATION FOR THE PRIOR DENSITY..
n
DATA> 7.7 .7
```

```
OBSERVED DATA IN WORKSHEET? (TYPE 'y' OR 'n'.)
   IF YES, INPUT NUMBER OF COLUMN.
   IF NO, INPUT OBSERVED SAMPLE MEAN, STANDARD DEVIATION, AND SAMPLE SIZE.
y
```

```
THE POSTERIOR DENSITY FOR M IS NORMAL
WITH MEAN AND STANDARD DEVIATION:
MEAN STD
   6.91923   0.25478
```

```
THE PREDICTIVE DENSITY OF THE NEXT OBSERVATION
IS NORMAL WITH MEAN AND STANDARD DEVIATION:
MEAN STD
   6.91923    0.81458
```

To summarize the distribution for the difference in means d, one runs the program 'mm_cont'. One inputs the mean and standard deviation of the normal density for each of the two population means M_1 and M_2 and the program gives the mean and standard deviation for the difference in means $M_1 - M_2$. In this case, $M_1 - M_2$ is approximately normal(0, .34), and so the difference $d = M_2 - M_1$ is also normal(0, .34). There appears to be little evidence from this data that quitting has any effect on average birthweight.

```
MTB > exec 'mm_cont'

INPUT MEAN AND STANDARD DEVIATION FOR NORMAL DISTRIBUTION FOR MEAN M1:
DATA> 6.92242    0.23484

INPUT MEAN AND STANDARD DEVIATION FOR NORMAL DISTRIBUTION FOR MEAN M2:
DATA> 6.91923    0.25478

THE POSTERIOR DENSITY FOR M1-M2 IS NORMAL
WITH MEAN AND STANDARD DEVIATION:
ROW          mn           st
  1    0.0031900    0.346501
```

7.3   Using t distributions and simulation

> Minitab command to learn about two means using continuous models
> (exact method):
> **exec 'mm_tt'**

The program 'mm_tt' describes an alternative method for obtaining the posterior distribution for the difference in means from two normal distributions. This program uses a more exact method than that implemented in 'mm_cont' in that it is based on the exact marginal posterior distribution for the means.

In this setting, there are four unknown parameters: the means M_1 and M_2 and the standard deviations S_1 and S_2 from the two populations. Since it is difficult in general to assess a joint prior distribution for four parameters, we assume that the parameters have the standard noninformative prior distribution proportional to $1/(S_1 S_2)$. Then the means M_1 and M_2 have independent t distributions. The posterior distribution of the difference in means $M_2 - M_1$ has a nonstandard functional form, but it is easy to obtain a simulated sample from this distribution. In this program, independent samples of a fixed size are simulated from the marginal posterior distributions of M_1 and M_2. If one pairs the two simulated samples and takes differences, simulated values of the posterior distribution of difference in means $d = M_2 - M_1$ are obtained.

This program is illustrated in the output here for the birthweight data. Unlike the previous run, we assume that little prior information exists about the locations of M_1 and M_2 and the corresponding population standard deviations, and therefore assign a noninformative prior distribution to the set of all parameters. We first indicate by typing "y" that the data sets are in the worksheet. The column locations of each data set are input, and one indicates that a simulated sample of size 1000 is desired. The output of the problem is a dotplot and summary statistics of the simulated values of the difference in means d. Comparing the outputs of 'mm_cont' and 'mm_tt', note that we obtain very similar inferences; the posterior distribution for $d = M_2 - M_1$ in both cases is approximately normal with mean 0 and standard deviation .36.

```
MTB > exec 'mm_tt'

OBSERVED DATA IN WORKSHEET? (TYPE 'y' OR 'n'.)
y

NOTE:
  INPUT NUMBER OF COLUMN OF FIRST DATASET
DATA> 1

NOTE:
  INPUT NUMBER OF COLUMN OF SECOND DATASET
DATA> 2

INPUT NUMBER OF SIMULATED VALUES:
DATA> 500
```

Simulated values of M1 and M2:
Each dot represents 3 points

```
                                      .
                                 .   :
                            .  .: :
                            :  ::::  .
                          ..: ::::::
                          ::::::::::::..
                         ::::::::::::::::..
                      ...:::::::::::::::::::.
                   :...::::::::::::::::::::::::... .      .
          ---+---------+---------+---------+---------+---------+---m1
```
Each dot represents 4 points
```
                                 .
                                 :
                            .  : .
                            :::::
                            :::::  .
                          .::::::::::::
                         :::::::::::::.:.
          .          .  .....::::::::::::::::::..   ...  . .
          ---+---------+---------+---------+---------+---------+---m2
           5.50      6.00      6.50      7.00      7.50      8.00
```

Simulated values of M2-M1:
Each dot represents 2 points
```
                               ..
                            :  :::::
                          .:.:.::::: :
                          .:::::::::::::
                         :::::::::::::::::.:
                         ::::::::::::::::::::
                     .   ::::::::::::::::::::: . :
              ....  ..:::.::::::::::::::::::::::.::.. ...        .
          ---+---------+---------+---------+---------+---------+---m_diff
           -1.00     -0.50      0.00      0.50      1.00      1.50
```

	N	MEAN	MEDIAN	TRMEAN	STDEV	SEMEAN
m_diff	500	-0.0195	-0.0134	-0.0177	0.3716	0.0166

	MIN	MAX	Q1	Q3
m_diff	-1.2028	1.5512	-0.2627	0.2173

7.4 Exercises

1. (Berry[2], exercise 12.25) A study was made to evaluate the effects of exercise and a nutrition supplement in very elderly people. A group of 83 people were randomly placed into exercise and nonexercise groups; the exercise group of 42 went through high-intensity resistance training of the knee and hip extensors for ten weeks, and the nonexercise group of 41 did not have this training. The measurement made was stair-climbing power (in percentage increase from a measurement made before the experiment). For the exercise group, the mean and standard deviation of stair-climbing power were 28.4 and 44; the values for the nonexercise group were 3.6 and 50. Using both the approximate method ('mm_cont') and the exact method ('mm_tt'), find a 95% posterior probability interval for the difference in population means $M_Y - M_N$, where M_Y is the average stair-climbing power of the exercise population and M_N is the mean of the nonexercise population. (Assume noninformative prior distributions for all unknown parameters.)

2. (Berry[2], exercise 12.29) An experiment compared the cortex weights of rats that lived in large, communal cages that had daily toy replacements with littermates that were isolated and had no toys. One is concerned about differences in the measurements for different experiments. For the control rats in experiment 1, the cortex weights were 657, 623, 652, 654, 658, 646, 600, 640, 605, 635, 642, and the cortex weights of the control rats in experiment 3 were 668, 667, 647, 693, 635, 644, 665, 689, 642, 673, 675, 641. Assuming flat priors for the corresponding population means, find the posterior probability that the control mean of experiment 3 is greater than that of experiment 1. Use both the exact and approximate methods.

3. (Lee[3], exercise 5.6) The following data consist of the lengths in millimeters of cuckoo's eggs found in nests belonging to the dunnock and to the reed warbler:

Dunnock	22.0	23.9	20.9	23.8	25.0	24.0	21.7	23.8	22.8	23.1
Reed warbler	23.2	22.0	22.2	21.2	21.6	21.9	22.0	22.9	22.8	

---

[2]Berry, D. (1996), *Statistics: A Bayesian Perspective*, Belmont, CA.: Duxbury Press.
[3]Lee, P. M. (1989), *Bayesian Statistics: An Introduction*, Oxford: Oxford University Press.

We are interested in estimating the difference $d = M_D - M_R$, where M_D (M_R) is the mean of the population of lengths of cuckoo's eggs belonging to the dunnock (reed warbler). Find a 90% posterior probability interval for this difference in means.

4. (Antleman[4], Chapter 10) Suppose a sample of fifty students who took the GMAT twice is selected. Twenty of the fifty took a GMAT preparatory course between their two attempts, and thirty did not. For those taking the preparatory course, the average improvement was fourteen points with a standard deviation of eight points; for those not taking the preparatory course, the average improvement was ten points with a standard deviation of seven points. Assume noninformative priors for the means.

   (a) Find the probability that the preparatory course helps.

   (b) Find the probability that the preparatory course adds at least five points to a person's first score.

---

[4]Antleman, G. (1996), *Elementary Bayesian Statistics*, Cheltenham: Edward Elgar Publishing.

Chapter 8

Learning about Relationships

8.1 Introduction

The programs described in this chapter are helpful in understanding relationships between two variables. Suppose that there are two measurements associated with each member of a population. The problem of interest is to learn about one of the measurements given a value of the second measurement. For example, suppose that applicants to a graduate program in mathematics take the Graduate Record Examination (GRE), which is divided into a verbal score and an analytical score. The population of interest is the group of all present and future applicants to the particular graduate program. One might think that there is a relationship between an applicant's verbal score and the analytical score. If a student does well on the verbal score, he or she is likely to perform well on the analytical part. If there is a relationship between the two variables, then the director of the graduate program may be able to use this information in prediction. A future applicant may take only the verbal part of the GRE. If the two scores are related, then this director will be able to use the student's verbal score to guess intelligently at the unknown analytical score.

In this illustration, the two measurements are continuous. One is also interested in studying relationships where the measurements are categorical. Larsen and Marx[1] describe the results of a market research study to learn about the relationship between an adult's self-perception and his or her attitude toward small cars. The population was

---

[1] Larsen, R. J. and Marx, M. L. (1990), *Statistics*, Englewood Cliffs, N.J.: Prentice-Hall.

people living in a specific metropolitan area. Each person in the sample was classified as favorable, neutral, or unfavorable toward small cars and, in addition, classified into one of three personality types. A car manufacturer is interested in looking at the relationship between personality type and like/dislike of small cars. If, for example, certain types of people like small cars, then the manufacturer could design its advertising to attract this particular segment of the population.

The Minitab program 'lin_reg' described in Section 8.2 is helpful for learning about the relationship between two continuous measurements. The focus is on two types of inferences. First, does a relationship exist? If so, one is interested in predicting a value of the second measurement given a known value of the first measurement. Section 8.3 illustrates the use of the program 'c_table' when the two measurements are categorical. A two-way contingency table is used to organize the measurements, and one wishes to test the hypothesis that there is an independence structure in the table. The program 'c_table' constructs a Bayesian test of the independence hypothesis, and this test result is contrasted with the classical test based on a chi-squared statistic.

8.2 Linear regression

> Minitab command for linear regression:
> **exec 'lin_reg'**

Consider the situation where two measurements x and y are made on each member of a random sample from a population. To illustrate this situation, we consider exercise 14.1 from Berry[2]. A study compared temperature of the air with the chirping frequency of the striped ground cricket. The observations are given in this table.

Chirps/sec:	20	16	20	18	17	16	15	17	15	16
Temp (° F):	89	72	93	84	81	75	70	82	69	83

Here the population can be thought as the temperature and chirping frequency for a large number of days, from which the ten paired measurements are a random sample. We distinguish the two measurements. The one that we want to learn about is called the *response variable*, and the one that will be useful in predicting the response is the *independent variable*. In this setting, we would like to use the chirping frequency to

---

[2]Berry, D. (1996), *Statistics: A Bayesian Perspective*, Belmont, CA.: Duxbury Press.

predict the temperature for a future day, so the temperature and frequency are the response and independent variables, respectively.

A simple model for this data is to assume that there exists a "true" straight-line relationship between the two measurements x and y of the form

$$y = a + bx,$$

where a is the intercept and b is the slope. If one were able to take measurements from every member of the population and plot the measurements on a scatterplot, then this "true" line would be a best line through the points. However, we don't know the intercept and slope for this true line, and we wish to learn about their values from the data in the sample.

Before we can learn about the true line, some additional assumptions are made. The values of the independent measurement x are viewed as constants, and, for a fixed value of x, the corresponding response y is assumed be random with a normal distribution with mean $a + bx$ and unknown standard deviation S. In our example, for a particular chirping frequency x, we assume that the temperature y can vary from day to day according to a normal distribution.

With these assumptions, we are interested in learning about all unknown quantities, the intercept a and slope b of the true line and the sampling standard deviation S from the sample of measurements. It is generally difficult to specify significant prior information for these three parameters, so a noninformative prior distribution will be assigned.

The program 'lin_reg' focuses on two types of inferences. We first are interested whether there is a significant relationship between the two variables. In our example, one may wonder whether chirping frequency has any relationship with the temperature. We address this problem by finding the marginal posterior density for the true line slope b. If a 95% probability interval for b does not include 0, then there is some evidence that a relationship does exist between the two variables. Next, if the two variables appear to be associated, then we are interested in using the independent variable to predict plausible values of the response. In general, we think of (x, y) as a future observation where the independent variable x is known and the response y is unknown, and we wish to *predict* the value of y. In our example, if tomorrow's chirping frequency is eighteen chirps/second, what is a good guess at the temperature? We will learn about this future temperature by summarizing the corresponding predictive distribution.

One complication in summarizing the marginal posterior and predictive densities is the fact that the standard deviation parameter S is unknown. However, Berry[3] has developed approximate normal approximations for these two densities that are easy to summarize. These approximations are used by the program 'lin_reg'.

To set up this problem, the data are placed in the Minitab worksheet (see the output). The values of the temperature are placed in column C1 named 'temp' and the frequency values are placed in column C2 named 'freq'. The program 'lin_reg' is now run. We input the numbers of the columns that contain the x (independent) and y (response) data.

```
MTB > name c1 'temp' c2 'freq'
MTB > set 'temp'
DATA> 89 72 93 84 81 75 70 82 69 83
DATA> end
MTB > set 'freq'
DATA> 20 16 20 18 17 16 15 17 15 16
DATA> end
MTB > exec 'lin_reg'

INPUT COLUMN NUMBERS OF X AND Y DATA:
DATA> 2 1

THE LEAST-SQUARES LINE HAS SLOPE AND INTERCEPT:
  ROW         B          A
   1    4.06665    10.6669

THE POSTERIOR DENSITY FOR b IS NORMAL
WITH MEAN AND STANDARD DEVIATION:
  ROW       MEAN         STD
   1    4.06665    0.669801

Input 'y' and 'return' to obtain prediction intervals:
y
NOTE:
  FOR PREDICTING Y FOR GIVEN VALUES OF X, INPUT X VALUES OF INTEREST:
DATA> 16 18 20
DATA> end
```

---

[3]Berry, D. (1996), *Statistics: A Bayesian Perspective*, Belmont, CA.: Duxbury Press.

NOTE:
    THE PREDICTIVE DENSITY OF THE NEXT OBSERVATION FOR DIFFERENT
    VALUES OF X IS NORMAL WITH MEANS AND STANDARD DEVIATIONS GIVEN BELOW:

    ROW   X_   MEAN_Y   STD_Y
     1   16   75.7334   3.90558
     2   18   83.8667   3.90558
     3   20   92.0000   4.34081

Input 'y' and 'return' to see graph of data with prediction intervals:
y

The program first gives the slope B and intercept A of the least-squares line. This is the line that best fits the ten sample points. Note that the slope of the line is 4.07. Is this significant? In other words, is this value far enough from 0 after allowing for sampling variability? This question is answered by looking at the posterior density of the slope of the true line slope b. This density is (approximately) normal with mean 4.07 and .67. A 95% posterior probability interval for b is $(4.07 - 1.96 \times .67, 3.66 + 4.07 \times .67) = (2.76, 5.38)$. This interval does not contain 0, so we can conclude that b is nonzero. There appears to be a significant relationship between the chirping frequency and the temperature.

We are interested next in predicting the temperature at the frequency values 16, 18, and 20.   We indicate by typing "y" that we wish to compute prediction intervals and place the values of the independent variable x on the DATA line. For each value of x that is input, the program gives the mean and standard deviation for the predicted y response. If tomorrow's chirping frequency is eighteen, then a 95% prediction interval for the temperature would be 83.9 plus and minus 1.95×3.9.

The program outputs a graph (see Figure 8.1) that summarizes the calculations. A scatterplot of the data is drawn. On top of the plot are drawn three lines: the middle line is the least-squares line, and the upper and lower lines are the bounds for 95% prediction intervals for chirping frequencies from fifteen to twenty. interval From this graph, one can quickly predict the temperature for any value within the range of the observed temperatures in the sample.

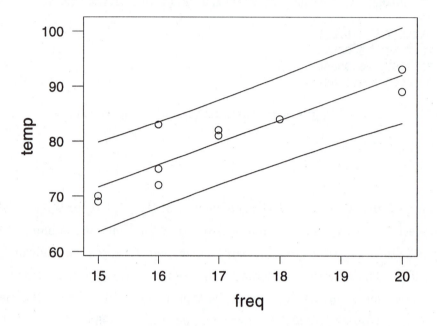

Figure 8.1: Least-squares line and prediction intervals for cricket example.

8.3   A test of independence of a contingency table

> Minitab command for testing for independence
> in a two-way contingency table:
> **exec 'c_table'**

A basic problem in statistics is to explore the relationship between two categorical measurements. To illustrate this situation, consider the following example presented in Moore[4] in which North Carolina State University looked at student performance in a course taken by chemical engineering majors. Researchers wished to learn about the relationship between the time spent in extracurricular activities and the grade in the course. Data on these two categorical variables were collected from 119 students, and the responses are presented using this contingency table:

---

[4]Moore, D. M. (1995), *The Basic Practice of Statistics*, New York: W. H. Freeman.

	Extracurricular activities (hours per week)		
	< 2	2 to 12	> 12
C or better	11	68	3
D or F	9	23	5

In this problem, the population of interest is all current and future chemical engineering majors who will take this particular course. One thinks that possibly the time spent in extracurricular activities has a detrimental effect on one's grade in this class.

To learn about the possible relationship between participation in extracurricular activities and grade, one tests the hypothesis of independence. This hypothesis says that the proportion of "C or better" students is the same for the three extracurricular activity groups. The usual non-Bayesian approach of testing the independence hypothesis first fits the independence model to the table and computes "expected" counts under this model for all the cells of the table. Then a test statistic is evaluated, which measures how far apart the expected counts are from the observed counts.

This test statistic is called the *Pearson chi-squared statistic*, and one rejects the hypothesis of independence if the value of the statistic is sufficiently large. One typically judges the size of the statistic by computing a *p value* that is the probability (under the hypothesis of independence) of observing a chi-squared statistic at least as large as the observed value. If this *p* value is sufficiently small, one rejects the independence hypothesis.

The program 'c_table' constructs an alternative Bayesian test of independence and contrasts this test with the non-Bayesian test described earlier that uses a chi-squared statistic and a *p* value. In this setting, there are two hypotheses. The hypothesis I is that the two categorical variables are independent, and the hypothesis D is that the two variables are dependent or related in some manner. From a Bayesian viewpoint, one makes a decision about these two hypotheses in terms of the Bayes factor that compares the probabilities of the observed data under the hypotheses:

$$BF = \frac{P(\text{data given D})}{P(\text{data given I})}.$$

This is referred as the Bayes factor in support of dependence or the Bayes factor against the independence hypothesis.

To compute a Bayes factor, one must construct two prior distributions: one under the independence hypothesis and one under the dependence hypothesis. We assume that the table in our example represents a random sample from the population of interest and the counts of the table have a multinomial distribution with proportion values given here:

	Extracurricular activities (hours per week)		
	< 2	2 to 12	> 12
C or better	p_{11}	p_{12}	p_{13}
D or F	p_{21}	p_{22}	p_{23}

Under the dependence hypothesis, the proportion values $p_{11}, \ldots, p_{23}$ can take any values that sum to 1, and the prior density places a uniform distribution over the set of all possible values. If the two variables are independent, then the proportions in the table can be determined by knowing the proportions of students in different activity levels and the proportions of students with different grades. For example, the proportion of students with high activity and who get C or better can be expressed as the product

$$p_{13} = P(\text{high activity})\, P(\text{C or better}).$$

In this independence case, the unknown parameters are the proportions of students in different activity levels and the proportions with different grades. We assume that our knowledge about these two sets of proportions are independent and assign to each set a uniform density over all possible values.

To use the program 'c_table', one first places the contingency table in a set of consecutive columns of the worksheet. In our example, the table has been placed in columns C1, C2, and C3. Then the program is run; the output is presented here. One inputs the number of the column where the table starts and the number of columns of the table. The program first presents the non-Bayesian chi-squared test of independence (using the Minitab "chisquare" command), and then gives the Bayes factor against the independence hypothesis.

```
MTB > exec 'c_table'

INPUT THE NUMBER OF THE FIRST COLUMN WHICH CONTAINS THE CONTINGENCY TABLE:
DATA> 1

INPUT THE NUMBER OF COLUMNS OF THE TABLE:
DATA> 3

Expected counts are printed below observed counts

           C1        C2        C3     Total
   1       11        68         3        82
         13.78     62.71      5.51

   2        9        23         5        37
          6.22     28.29      2.49

Total      20        91         8       119

ChiSq =  0.561 +  0.447 +  1.145 +
         1.244 +  0.991 +  2.538 = 6.926
df = 2, p = 0.032
1 cells with expected counts less than 5.0

----------------------------------------------------------
The Bayes factor against the hypothesis of independence
with uniform priors is:
BAYES_F
  1.66221
```

For these data, the value of the chi-squared statistic is 6.926. The p value is approximately .03. Since this value is smaller than 5% (the popular testing probability), it can be viewed (from a non-Bayesian viewpoint) as significant evidence that activity and grade levels are not independent.

How does this result compare with the Bayesian test? The value of the Bayes factor is 1.66, which means that the dependence hypothesis D is approximately 1.5 times as likely as the independence hypothesis. If the prior probability of independence is given by q, then the posterior probability of independence is given by

$$P(\text{H given data}) = \frac{1-q}{qBF + 1 - q},$$

where BF is the value of the Bayes factor in support of dependence. In this case, if one is uncertain initially what hypothesis was true, the probability q can be set to .5, and the posterior probability of independence is

$$P(\text{H given data}) = \frac{.5}{.5 \times 1.66 + .5} = .376.$$

In contrast to the conclusion of the non-Bayesian test, the Bayesian test would probably not reject independence since .376 is not a small probability.

The non-Bayesian and Bayesian procedures also display different behavior for testing independence in tables with different sample sizes. To illustrate, consider the following hypothetical contingency table that is shown in the following output. The total sample size is twice that of the first table, but the chi-squared value is approximately the same value (7.1). Again the p value is equal to 3%, and the non-Bayesian test would conclude that the two variables were not independent. In the output, we note that the value of the Bayes factor in support of dependence is equal to .89. Thus the Bayes test actually prefers the independence hypothesis. This illustrates one difficulty of interpreting p values. A p value of 3% for a sample of size 238 is actually less evidence against independence than the same p value for a sample of size 119. The Bayesian test procedures, in contrast, do properly account for this sample size effect.

```
MTB > exec 'c_table'

INPUT THE NUMBER OF THE FIRST COLUMN WHICH CONTAINS THE CONTINGENCY TABLE:
DATA> 4

INPUT THE NUMBER OF COLUMNS OF THE TABLE:
DATA> 3

Expected counts are printed below observed counts

          C4        C5        C6      Total
    1      22       138        11       171
         28.74    130.76     11.50

    2      18        44         5        67
         11.26     51.24      4.50

Total     40       182        16       238
```

```
ChiSq =  1.580 +  0.400 +  0.021 +
         4.034 +  1.022 +  0.055 = 7.112
df = 2, p = 0.029
```

```
The Bayes factor against the hypothesis of independence
with uniform priors is:
BAYES_F
  0.889744
```

8.4 Exercises

1. (Berry[5], example 14.1) A study compared two drugs — formoterol and salbutamol — in treating patients suffering from exercise-induced asthma. The measurement of interest is forced expiratory volumes (in milliliters). Thirty patients used both drugs with the measurements recorded in this table. Let F denote the response (using the drug formoterol) and S the independent variable.

Patient	F	S	Patient	F	S
1	35	32	16	34	24
2	31	24	17	23	22
3	32	21	18	14	16
4	26	22	19	23	22
5	29	19	20	13	15
6	25	24	21	23	13
7	20	26	22	38	26
8	28	28	23	30	24
9	32	33	24	32	31
10	34	28	35	30	26
11	28	22	26	26	24
12	23	16	27	31	28
13	14	10	28	31	28
14	31	14	30	28	27
15	27	22	31	25	22

(a) Is the salbutamol measurement a useful predictor of a patient's formoterol measurement? (Look at the posterior density of the true slope parameter b.)

---

[5]Berry, D. (1996), *Statistics: A Bayesian Perspective*, Belmont, CA.: Duxbury Press.

(b) If a future patient's measurement using the drug salbutanmol is thirty, find a 90% prediction interval for his or her measurement using the drug formoterol. Also find a 90% prediction interval if the patient's salbutanmol measurement is twenty.

2. (Berry[6], exercise 14.6) This table gives the number of home and away wins for twenty-six major league baseball teams in 1992. Let the number of home wins be the response variable and the number of away wins be the independent variable.

Home	Away	Home	Away	Home	Away	Home	Away
51	47	41	35	43	44	45	37
43	46	38	37	41	31	42	30
44	29	47	34	41	35	38	26
41	31	44	28	51	45	36	41
43	35	37	26	41	29	53	43
50	36	53	39	53	43		
53	37	48	42	45	38		

(a) By looking at a posterior probability interval for b, test the claim that the number of home wins of a team is unrelated to the number of away wins.

(b) If a team in a future season wins fifty away games during a season, find the predictive probability that the team will win a total of one hundred games or more. Find this probability if the team only wins forty away games during the season.

3. (Andersen[7]) In 1985 Radio Denmark conducted a survey regarding the interest among TV viewers with the Saturday afternoon broadcast called *Sportslordag*. A group of 635 people said they had seen *Sportslordag* at least once. They were asked about their preferences about the length of the broadcast. The answers were grouped in four time intervals. In addition, the people were classified with respect to their sex and age. The table of counts is given here. Test the hypothesis that one's wishes to the length of *Sportslordag* are independent of one's age and sex using a Bayes factor. Compare this measure of evidence with the p value of a classical chi-squared test.

---

[6] Berry, D. (1996), *Statistics: A Bayesian Perspective*, Belmont, CA.: Duxbury Press.
[7] Andersen, E. B. (1990), *The Statistical Analysis of Categorical Data*, New York: Springer-Verlag.

Age/sex stratum	Wishes to the length of *Sportslordag*			
	Less than 2 hours	2.5 to 3.5 hours	4 hours or more	Do not know
Men, above 40	65	63	59	5
Women, above 40	77	39	32	4
Men, under 40	81	50	30	2
Women, under 40	80	38	6	4

4. (Andersen[8]) To study the effects of air pollution, the number of lung cancer cases was observed for the years 1968 to 1971 for four Danish cities. In the following table, the cases are distributed over six age groups. One wishes to test the hypothesis that the distribution of lung cancer cases for different ages is the same for each city. Test this hypothesis of independence of age and city by means of a Bayes factor and a classical chi-squared statistic. Compare these two measures of evidence.

Age	City			
	Fredericia	Horsens	Kolding	Vejle
40-54	11	13	4	5
55-59	11	6	8	7
60-64	11	15	7	10
65-69	10	10	11	14
70-74	11	12	9	8
Over 75	10	2	12	7

---

[8]Andersen, E. B. (1990), *The Statistical Analysis of Categorical Data*, New York: Springer-Verlag.

Chapter 9

Learning about Discrete Models

9.1 Introduction

> Minitab command to learn from discrete models:
> **exec 'mod_disc'**
> Minitab command to summarize a discrete probability distribution:
> **exec 'disc_sum'**

Bayes' rule for a discrete collection of models was introduced in Chapter 3 and illustrated for binomial sampling in Chapters 4 and 5 and normal sampling in Chapter 6. The program 'mod_disc' implements a general method for learning about a single unknown parameter from a specified distribution. This parameter could be the proportion p from a binomial sample, the mean M from a normal population with known standard deviation, or parameters from other distributions. The program 'mod_disc' is set up to perform inference about binomial, normal, Poisson, hypergeometric, exponential, uniform, and capture-recapture populations, and the program can easily be modified to learn about other distributions. The output of this program is a discrete probability distribution that reflects one's opinion about the parameter after observing data. The macro 'disc_sum' can be used to summarize this probability distribution.

To illustrate this general approach for learning, consider inference about the mean parameter of a Poisson distribution. Suppose that a manager of a car dealership is interested in estimating the daily rate at which a particular new salesman sells cars. What the manager doesn't know is the daily rate L, which is the average number of cars that the salesman would sell in a hypothetical infinite number of days. We will refer to this unknown rate as a *model*, and the manager will learn about the probabilities of different models by observing how many cars the salesman sells in a twenty-four-day period.

The first step in the learning process is for the manager to construct a prior distribution for the unknown model L that reflects his opinion about the selling ability of the car salesman. The manager first thinks of a set of possible values for the model (selling rate) and assigns probabilities to each model that reflect his opinion about the plausibility of these different rates. Suppose that the manager wishes to classify the salesman as "good," "average," or "poor," and these categories correspond to the model values .5, .25, and .125, respectively. Based on his knowledge about salesmen of comparable experience, he believes that L takes on these three values with probabilities .2, .5, and .3, respectively. Here these probabilities can be viewed as relative frequencies; of all the salesmen who have worked for the dealership with similar experience, 20% of them have been good salesmen, 50% have been average, and 30% have been poor.

Next, the manager observes the salesman's performance during a twenty-four-day period. The salesman sells ten cars during this time. What can the manager conclude about the salesman's ability to sell cars?

The manager can update his probabilities about the models by means of Bayes' rule. The calculations are performed in the familiar format that is displayed in the output here. The first column, MODEL, lists the possible selling rates, and the column PRIOR lists the corresponding initial probabilities of the models. The next column, LIKE, lists the likelihood that is the probability of the data (ten cars sold in twenty-four days) for each model value. Here we assume that selling a car is a relatively rare event, and so the number of cars sold in twenty-four days is approximately Poisson distributed with mean 24 L, where L is the daily selling rate. In this case, the probability of selling ten cars in twenty-four days, for a given rate L is proportional to

$$LIKE = \exp(-24L)(24L)^{10}.$$

The likelihood values presented in the following table have been scaled for readability — the largest value of the likelihood is given by 1000000. Once the prior probabilities and the likelihood values are given, the updated probabilities are found by multiplying the probabilities by the likelihoods to get products (shown in the column PRODUCT), and these products are divided by their sum to get posterior probabilities in the column POST.

```
ROW   MODEL   PRIOR      LIKE   PRODUCT       POST
  1   0.500     0.2   1000000    200000   0.500870
  2   0.250     0.5    393974    196987   0.493324
  3   0.125     0.3      7728      2318   0.005806
```

What does the manager now think about the salesman's ability? The posterior probabilities are roughly equally divided between $L = .5$ (good) and $L = .25$ (average). The manager is pretty confident that the salesman is at least average, and so the salesman probably won't lose his job in the near future.

The program 'mod_disc' is set up to learn about parameters from seven different sampling distributions. The seven types of likelihoods, the corresponding model or parameter of interest, and the associated data are listed in the following table. For a binomial problem, the model is the proportion p, and the data are the number of successes and failures in the binomial experiment. The binomial proportion and normal mean problems are included in collection of likelihoods, so this program can perform the same posterior calculations as the programs 'p_disc' and 'm_disc' discussed in Chapters 4 and 6.

LIKELIHOOD	MODEL	DATA
Binomial	p	# successes, # failures
Normal	M	sample mean, sample size, pop. std. dev.
Poisson	L	sample sum, time interval
Hypergeometric	S	pop. size, sample size, # successes
Discrete uniform	N	maximum obs., sample size
Capture-recapture	N	# marked, sample size, # marked in sample
Exponential	M	sample sum, sample size

Before the program is run, two Minitab columns must be defined. The possible values of the model (parameter) are placed in the column 'MODEL' and the corresponding prior probabilities in the column 'PRIOR'. For our example, 'MODEL' would contain the three possible selling rates and 'PRIOR' the initial probabilities of these three rates.

After the columns 'MODEL' and 'PRIOR' are defined, the program 'mod_disc' is run. There are two inputs: the definition of the likelihood and the corresponding data. One is given a choice of seven likelihood definitions and inputs the number of the likelihood desired. The program will explain what data should be entered for that particular definition of likelihood, and the user will input the data values on the DATA line.

The program will output the updated posterior probabilities in a table format. The values of the posterior probabilities are stored in the column 'POST' and plotted using a vertical line graph. To quickly summarize these probabilities, the means of the prior and posterior probabilities are printed. These means are useful in seeing how one's probabilities are changed by the observed data.

In the following sections, we illustrate the generality of this modeling approach by considering inference for some nonstandard distributions. We first show the Minitab output for the Poisson mean example discussed earlier. Then we talk about learning about the unknown upper bound of an uniform distribution and the unknown number of successes of a finite population (hypergeometric sampling).

The final section of this chapter uses a Minitab program to introduce the notion of model criticism. It can be difficult for you to specify initial probabilities for a set of models, and the probabilities that are used should be viewed as an approximation to your "true probabilities" if you were able to think about the problem for a long period of time. Thus more than one set of probabilities may exist that may represent your prior belief about a parameter. In this case, there are two concerns. Do the posterior probabilities depend significantly on the choice of prior? If they do not, then the posterior probabilities are *robust* to the choice of prior. If different priors do give different posterior probabilities, then is one set of prior probabilities preferable or more consistent with the observed data? One can answer this question by means of a Bayes factor that compares the probability of the data under the first prior model with the probability of the data under the second model. The program 'mod_crit' performs these "model criticism" computations for the general situation where one is learning about a discrete set of models. This program is illustrated in Section 9.5 for a proportion problem.

9.2 Inference about a Poisson mean

The problem of learning about the unknown selling rate of a car salesman was discussed in the introductory section. The full output of the program 'mod_disc' is shown below for this example. This program is set up by defining two columns 'model' and 'prior'. Then in the run of 'mod_disc', we input "3" to indicate that we are using a Poisson likelihood and enter the sample sum (total number of sales) and time interval (number of days) on the DATA line. The posterior probabilities are listed in the table and stored in the column 'POST'. A graph of these probabilities is given in Figure 9.1.

```
MTB > name c1 'model' c2 'prior'
MTB > set 'model'
DATA> .5 .25 .125
DATA> end
MTB > set 'prior'
DATA> .2 .5 .3
DATA> end
MTB > exec 'mod_disc'

INPUT THE NUMBER OF THE LIKELIHOOD:
(1-Binomial P, 2-Normal M, 3-Poisson L, 4-Hypergeometric S,
 5-Discrete Uniform N, 6-Capture/Recapture N, 7-Exponential M)
DATA> 3

INPUT
  (sample sum, time interval)
DATA> 10 24
```

ROW	model	prior	LIKE	PRODUCT	POST
1	0.500	0.2	1000000	200000	0.500870
2	0.250	0.5	393974	196987	0.493324
3	0.125	0.3	7728	2318	0.005806

```
PRIOR MEAN OF MODELS:
MEAN
  0.2625

POSTERIOR MEAN OF MODELS:
MEAN
  0.374492
```

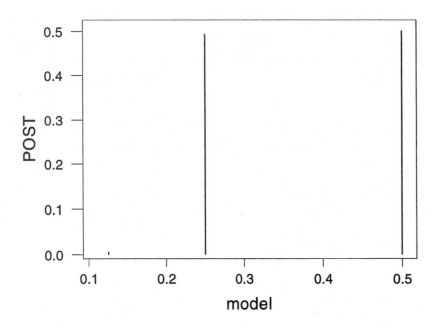

Figure 9.1: Posterior probabilities of models for Poisson example.

9.3   Inference about the upper bound of a population

Suppose that one takes independent observations from a uniform distribution on the set $\{1, 2, \ldots, N\}$, where the total number of possible values N is unknown. To illustrate this situation, suppose that a tourist is waiting for a taxi in a city. During this waiting time, she observes the numbers of five taxis that drive by. Can she estimate the total number of taxis in the city from this information?

To learn about the unknown total number, some assumptions need to be made. First, the tourist will assume that the taxis in the city are numbered from 1 to some unknown number N. Second, at a given time, she is equally likely to observe any particular taxi (each taxi has a route through the entire city) and repeated observations are independent. Finally, the tourist may have some prior knowledge about N since she has some familiarity with the size of the city. In this particular application, we will assume that this prior knowledge is weak and only assume that N is equally likely to be any positive integer from 1 to a maximum value B that is supplied by the tourist. In this example, we suppose that the tourist thinks that there cannot be more than 200 taxis in the city, so $B = 200$.

Now the tourist observes five taxis with numbers 43, 24, 100, 35, and 85. What has she learned about the total number of taxis in the city?

The program 'mod_disc' can be used to summarize the posterior distribution for the unknown number of taxis N (see the output). To set up this problem, one names two columns 'model' and 'prior'. The possible values for N are placed in the column 'model' — here the models consist of all numbers from 1 to 200. The associated probabilities of the models are placed in the column 'prior'; since all 200 models are assigned the same prior, each model receives a probability of $1/200 = .005$.

The macro 'mod_disc' is run next. One inputs the number 5 to indicate that a discrete uniform likelihood is to be used. For this problem the data are the maximum observation and the sample size – here the largest taxi number was 100, and five observations were taken. The output is the list of all possible values for N and the corresponding posterior probabilities. These updated probabilities are stored in the column 'POST'.

```
MTB > name c1 'model' c2 'prior'
MTB > set 'model'
DATA> 1:200
DATA> end
MTB > let 'prior'=.005+0*'model'
MTB > exec 'mod_disc'

INPUT THE NUMBER OF THE LIKELIHOOD:
(1-Binomial P, 2-Normal M, 3-Poisson L, 4-Hypergeometric S,
 5-Discrete Uniform N, 6-Capture/Recapture N, 7-Exponential M)
DATA> 5

INPUT
  (maximum observation, sample size)
DATA> 100 5
```

Row	model	prior	LIKE	PRODUCT	POST
1	1	0.005	0	0.00	0.0000000

2	2	0.005	0	0.00	0.0000000
3	3	0.005	0	0.00	0.0000000
198	198	0.005	32861	164.30	0.0013716
199	199	0.005	32043	160.22	0.0013375
200	200	0.005	31250	156.25	0.0013044

PRIOR MEAN OF MODELS:

MEAN
  100.5

POSTERIOR MEAN OF MODELS:

MEAN
  123.976

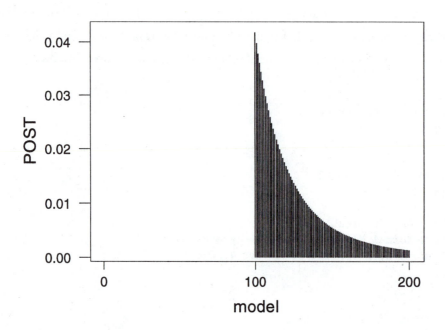

Figure 9.2: Posterior probabilities of models for discrete uniform example.

Based on the computations made on the worksheet and the accompanying graph (Figure 9.2), various types of inferences about N can be made. The most likely value of N, the posterior mode, is clearly 100 from the graph. The posterior mean of N given in the output is 123.976. (The fact that the posterior mean is larger than the posterior mode is a reflection of the right skewness of the distribution of posterior probabilities.) One can check this calculation of the posterior mean using the model values in 'model' and the updated probabilities in 'POST'. The mean is found by multiplying these two columns and summing the result:

```
MTB > let k1=sum('model'*'POST')
MTB > print k1
K1       123.976
```

The program 'disc_sum', described in Section 4.2.2, is helpful for computing probabilities and for finding probability intervals for the discrete distribution for N. A portion of the output of the run of this program is displayed here. The input to the program is the numbers of the columns that contain the values of the model and the associated probabilities.

Suppose that one is interested in the probability that the number of taxis exceeds 150. In the output of the program 'disc_sum', note that the cumulative probability that N is less than or equal to 150 is .861. Thus, the probability that N exceeds 150 is $1 - .861 = .139$. Also the program finds a set of values of N that has a probability content exceeding .9. The probability that N is in the set $\{100, 101, ..., 159\}$ is equal to .903.

```
MTB > exec 'disc_sum'

INPUT NUMBER OF COLUMN WHICH CONTAINS VALUES OF VARIABLE:
DATA> 1

INPUT NUMBER OF COLUMN WHICH CONTAINS PROBABILITIES:
DATA> 51

TYPE 'y' TO COMPUTE CUMULATIVE PROBABILITIES:
y
DATA> 150
DATA> end
```

```
Row  VALUE    PROB_LE
  1    150    0.861175
```

TYPE 'y' TO COMPUTE PROBABILITY INTERVALS:
```
y
DATA> .9
DATA> end
```

```
PROB_SET
  0.90334
```

```
SET
     100    101    102    103    104    105    106    107    108    109    110
     111    112    113    114    115    116    117    118    119    120    121
     122    123    124    125    126    127    128    129    130    131    132
     133    134    135    136    137    138    139    140    141    142    143
     144    145    146    147    148    149    150    151    152    153    154
     155    156    157    158    159
```

9.4   Inference about a finite population

In Chapter 4, the problem of learning about a population proportion p was considered. We assumed that the population consisted of two types (successes and failures) and the proportion of successes in the population was p. A random sample was selected from the population, and s successes and f failures were observed. The goal was to learn about the proportion p from this data.

This scenario implicitly assumes that the size of the population is much larger than the size of the sample. Then a sample from this population can be regarded (approximately) as taken *with replacement*. The computation of the likelihood in Chapter 4 made the assumption that, after taking an observation (either success or failure), it is placed back into the population before the next observation is selected.

This assumption that the observations are taken with replacement will be inappropriate when the size of the sample is relatively large in comparison to the size of the population. We will call this a *finite population* problem. In that case, the random sample should be regarded as being taken *without replacement* and the likelihood function will be given by a hypergeometric formula.

In general, suppose that a finite population of known size N consists of successes and failures. Suppose that a random sample of size n is taken without replacement, and x successes are observed. The problem is to learn about the unknown number K of successes in the population. Equivalently, one is interested in learning about the proportion of successes $p = K/N$.

As an example, suppose that a small community has 100 voters and you are interested in estimating the number of voters in favor of the proposed school tax levy. You take a random sample of twenty voters and twelve are in favor. What have you learned about the number of voters in favor of the levy in the community?

Before proceeding, we consider possible prior distributions for the number of successes in the population K. In practice, one has some opinion about likely values of K. For example, one is probably certain that the number of successes is larger than 0 and smaller than the population size. However, in this example, we will suppose that K takes on all possible values with equal probability. This reflects vague or imprecise prior beliefs about the number of successes. Also, if a large sample is taken, the prior distribution will have a relatively small impact on the posterior distribution. In this case, this uniform prior distribution will be a suitable approximation to other informative priors that can be chosen for K.

The program 'mod_disc' is used to learn about K (see the output). For this problem, a model is a value of the number of successes K. The possible values for K (values from 0 to 100) are placed in the column 'model', and the corresponding initial probabilities are placed in the column 'prior'. When one runs the program, one inputs 4 to indicate that a hypergeometric likelihood will be used. The data for this problem are the known population size, the sample size, and the number of successes in the sample. In this example, the values are 100, 20 and 12. The program lists the 101 possible values of K and the corresponding prior and posterior probabilities. (The posterior probabilities are stored in the column 'POST'.)

```
MTB > name c1 'model' c2 'prior'
MTB > set 'model'
DATA> 0:100
DATA> end
MTB > let 'prior'=0*'model'+1/101
MTB > exec 'mod_disc'
```

```
INPUT THE NUMBER OF THE LIKELIHOOD:
(1-Binomial P, 2-Normal M, 3-Poisson L, 4-Hypergeometric S,
 5-Discrete Uniform N, 6-Capture/Recapture N, 7-Exponential M)
DATA> 4

INPUT
  (population size, sample size, number of successes)
DATA> 100 20 12
```

ROW	model	prior	LIKE	PRODUCT	POST
1	0	0.0099010	0	0.00	0.0000000
2	1	0.0099010	0	0.00	0.0000000
3	2	0.0099010	0	0.00	0.0000000
56	55	0.0099010	878708	8700.08	0.0366777
57	56	0.0099010	919530	9104.26	0.0383817
58	57	0.0099010	952999	9435.63	0.0397787
59	58	0.0099010	978043	9683.59	0.0408240
60	59	0.0099010	993923	9840.82	0.0414869
61	60	0.0099010	1000000	9900.99	0.0417405
62	61	0.0099010	995919	9860.58	0.0415702
63	62	0.0099010	981639	9719.20	0.0409741
64	63	0.0099010	957364	9478.85	0.0399609
65	64	0.0099010	923559	9144.14	0.0385498
97	96	0.0099010	0	0.00	0.0000000
98	97	0.0099010	0	0.00	0.0000000
99	98	0.0099010	0	0.00	0.0000000
100	99	0.0099010	0	0.00	0.0000000
101	100	0.0099010	0	0.00	0.0000000

```
PRIOR MEAN OF MODELS:
MEAN
  50

POSTERIOR MEAN OF MODELS:
MEAN
  59.2748
```

The plot of the posterior probabilities is shown in Figure 9.3. The posterior mode, the value of K with the largest posterior probability, is 60. One can find an interval estimate using the stored posterior probabilities and the macro 'disc_sum'. To illustrate,

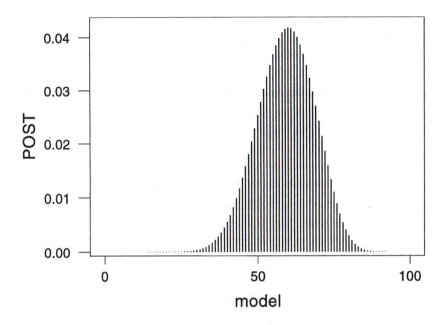

Figure 9.3: Posterior probabilities of models for finite population example.

suppose that we wish to find an interval of K values that contains at least 90% of the posterior distribution. We run the program 'disc_sum'. The numbers of the columns that contain the probability distribution are entered and we indicate that we wish to find a probability interval of probability content .9. The set of values {44, 45, ..., 74} is output from this macro. So the interval [44, 74] is an approximate 90% probability interval for the unknown number of successes K. In other words, since the population size is 100, we believe that the population proportion p lies between .44 and .74 with probability .90.

9.5   Model criticism: comparing priors

> Minitab command to compare two priors from discrete models:
> **exec 'mod_crit'**

In the examples of this chapter, a prior distribution was constructed, and the focus was on the learning from the set of posterior probabilities. But it may be difficult to specify probabilities, and several prior distributions may exist that may serve as reasonable approximations to one's prior opinion about a parameter. The program 'mod_crit' can be used in the case where there are two possible sets of prior probabilities. This program will compute the posterior probabilities for each prior; by comparing these two distributions, one can see whether the exact choice of prior does have an impact on the posterior distribution. Also, by use of a Bayes factor, the program will indicate which prior distribution is a better fit to the observed data.

To illustrate the comparison of priors, we will consider a baseball example similar to the one used in Chapter 4. Suppose that a manager is evaluating a new baseball player. He wishes to learn about the player's true batting average p. Suppose that this manager gets information from two scouts who have watched the player play in the minor leagues. The first scout thinks that the player is an average player and will bat around .270. But he's not sure — it's possible that the player will excel and bat as high as .340 or have a poor season and bat only .200. The second scout has a lower opinion about the player. He thinks that the same set of batting averages are possible but places a much greater weight on the smaller averages. The prior distributions for the two scouts are given in this table.

p	.200	.220	.240	.260	.280	.300	.320	.340
Prior 1	.05	.05	.10	.25	.25	.15	.10	.05
Prior 2	.20	.20	.20	.15	.10	.05	.05	.05

Now suppose that the player plays for a week and gets ten hits in his first thirty at-bats. Both scouts, with these new data, can update their probabilities. In this case, is the choice of prior important? Do the scouts still have different opinions about the player after seeing these data? And if they do have different opinions, which scout had an opinion that was more consistent with the hitter's performance?

The program 'mod_crit' is helpful in comparing and criticizing two possible priors for a parameter with a discrete prior distribution. To set up this program, three Minitab columns are defined. The column 'model' contains the model (parameter) values, and the columns 'prior1' and 'prior2' contain the two sets of prior probabilities. For our baseball example, the values of the proportion p and the prior probabilities for the two scouts are placed in these three columns.

The program 'mod_crit' is next run (see the output below). One inputs the particular likelihood (binomial in this case) and the data (ten successes and twenty failures where a success is defined as a hit). The output is a table that lists the models, the two prior distributions that were entered, and the two updated posterior distributions. To help understand the difference between the two sets of posterior probabilities, the respective prior and posterior means are given. The first scout initially had a twenty-seven-point $(.274 - .247)$ higher opinion of the player. After seeing the data, the first scout has only a eighteen-point $(.283 - .265)$ higher opinion. So the beliefs of the two scouts are more similar after seeing these hitting data.

```
MTB > exec 'mod_crit'

INPUT THE NUMBER OF THE LIKELIHOOD:
(1-Binomial P, 2-Normal M, 3-Poisson L, 4-Hypergeometric S,
 5-Discrete Uniform N, 6-Capture/Recapture N, 7-Exponential M)
DATA> 1

INPUT
  (number of successes, number of failures)
DATA> 10 20
```

Row	model	prior1	prior2	POST1	POST2
1	0.20	0.05	0.20	0.015655	0.084360
2	0.22	0.05	0.20	0.024472	0.131873
3	0.24	0.10	0.20	0.069496	0.187250
4	0.26	0.25	0.15	0.226925	0.183429
5	0.28	0.25	0.10	0.275262	0.148333
6	0.30	0.15	0.05	0.187430	0.084169
7	0.32	0.10	0.05	0.133430	0.089878
8	0.34	0.05	0.05	0.067331	0.090708

```
TYPE 'y' AND RETURN FOR SUMMARIES:
y
```

```
FOR FIRST SET OF PRIOR PROBABILITIES:
---------------------------------------
PRIOR MEAN OF MODELS:
MEAN
  0.274

POSTERIOR MEAN OF MODELS:
MEAN
  0.283087

FOR SECOND SET OF PRIOR PROBABILITIES:
---------------------------------------
PRIOR MEAN OF MODELS:
MEAN
  0.247

POSTERIOR MEAN OF MODELS:
MEAN
  0.264901


-----------------------------------------------------------
BAYES FACTOR IN FAVOR OF FIRST SET OF PRIOR PROBABILITIES:
BAYES_F
  1.34720

TYPE 'y' AND RETURN TO SEE PLOTS:
y
```

To criticize the two sets of probabilities, one computes the Bayes factor, which is the ratio of the probability of the data under the two priors:

$$BF = \frac{P(\text{data given Prior 1})}{P(\text{data given Prior 2})}.$$

From the output, we see that the Bayes factor in support of the first scout's prior is 1.35. This indicates that the first prior distribution is slightly more consistent with the observed data. The first scout had a higher initial opinion of the player that was more consistent with the player's ten hits in thirty at-bats.

Let's return to the manager who has to make a decision about the batter's ability. Suppose that his knowledge of the batter is based solely on the information provided by the two scouts. How does he update his probabilities? Suppose that he has equal confidence initially in the two scouts' knowledge, and so his prior probability of the model p is given by the mixture

$$\frac{1}{2}\pi_1(p) + \frac{1}{2}\pi_2(p),$$

where $\pi_1(p)$ and $\pi_2(p)$ denote the prior probabilities of the proportion value for the two scouts. After observing the hitting data, the manager's updated probability of the proportion is given by the new mixture

$$\frac{BF}{BF+1}\pi_1(p \mid \text{data}) + \frac{1}{BF+1}\pi_2(p \mid \text{data}),$$

where $\pi_1(p \mid \text{data})$ and $\pi_2(p \mid \text{data})$ are the scouts' posterior probabilities, and BF is the Bayes factor defined above.

These posterior probabilities of the manager can be computed from the output to 'mod_crit'. The first set of posterior probabilities are stored in the column 'post_1' and the second set in 'post_2'. The value of the Bayes factor is stored in the column 'bayes_f'. Then, if a new column is named 'post', the manager's probabilities are found by the Minitab let command:

```
let 'post' = 'bayes_f'/('bayes_f'+1)*'post_1'+1/('bayes_f'+1)*'post_2'
```

In this example, the manager's posterior probabilities are a mixture of .574 times the first scout's probabilities and .426 times the second scout's probabilities. He now places slightly higher confidence in the first scout's opinion. Also, his opinion about the player's ability is a weighted average of the two scout's beliefs, with a larger weight assigned to the first scout.

9.6 Exercises

1. (Inference about an exponential mean) You are interested in the mean time that a particular type of light bulb burns. Five bulbs are tested with observed burn times (in hours) of 751, 594, 1213, 1126, 819. Assume that each burn time is distributed exponential with mean M. If the observations are a random sample from this distribution, then the likelihood function is given by

$$l(M) = M^{-n}\exp\{-s/M\},$$

where n is the size of the sample and s is the sum of the observations. Suppose that M can take on the values 500, 1000, 1500, and 2000 with equal probabilities. Use the program 'mod_disc' to find the posterior probabilities of M. (In the input of the program specify an exponential likelihood. The data column consists of the sum s and the sample size n.)

2. (Inference about the upper bound of a discrete uniform population) Suppose that a bowl contains balls that are numbered from 1 to an unknown value N. To learn about N, you sample five balls with replacement from the bowl — suppose the numbers you select are 20, 43, 80, 54, 76.

   (a) Before sampling, suppose that you believe that the number of balls in the bowl is equally likely to be any number from 50 to 300. Find the posterior distribution for N and find an interval that contains 90% of the posterior probability.

   (b) Consider the use of a more informative prior distribution. Suppose that you think that the prior distribution for N is given by this table.

N	50	100	150	200	250
$P(N)$	.1	.2	.4	.2	.1

   Find the posterior distribution for N. Compare the inferences using the two priors by computing the mean and posterior standard deviation of each posterior distribution. How influential is the informative prior distribution in this case?

3. (Inference about the number of successes in a finite population) In a lot of thirty unopened boxes of word processing software, you are interested in learning about the number N that don't include the full set of documentation. Before sampling, you have little knowledge about the value of N and so you assign the values 0, 1, ..., 30 equal prior probabilities. You are reluctant to open all thirty boxes. So instead you take a random sample of ten boxes with replacement from the lot and find that three of the software boxes are missing some documentation. Find the posterior probabilities for N. Find the value of N that is most likely. In addition, find the probability that over half of the boxes in the lot are missing some documentation.

4. (Inference about the size of a population using capture/recapture sampling) You wish to estimate the number of fish N in a lake. Suppose, before sampling, you believe that the possible values of N are 1000, 2000, 3000, 4000, 5000, and each value is equally likely. You obtain more information about the population size using a capture/recapture sampling method. You introduce 200 marked fish in the lake. After a week, you take a sample of 100 with replacement from the lake and find 32 marked fish. Use 'mod_disc' to find the posterior distribution for N. (The capture/recapture likelihood is used and the data consists of the number of marked fish introduced, the sample size, and the number of marked fish in the sample.)

5. (Antleman[1], Chapter 8) Suppose you own a trucking company with a large fleet of trucks. Breakdowns occur randomly in time and the number of breakdowns during an interval of t days is assumed to be Poisson distributed with mean tL. The parameter L is the daily breakdown rate. The possible values for L are .5, 1, 1.5, 2, 2.5, 3 with respective probabilities .1, .2, .3, .2, .15, .05.

   (a) If twelve trucks break down in a six-day period, find the posterior probabilities for the different rate values.

   (b) Find the probability that there are no breakdowns during the next week.

      Hint: If the rate is L, the conditional probability of no breakdowns during a seven-day period is given by $\exp\{-7L\}$. One can compute this probability on Minitab by multiplying the column of conditional probabilities by the posterior probabilities of L and finding the sum of the products.

      ```
      let k1=sum(exp(-7*'model')*'post')
      print k1
      ```

6. (Berry[2], exercise 6.15) A study was designed to see whether there would be fewer injuries in baseball with breakaway bases. In a large number of games with breakaway (B) bases there were two sliding injuries. In about the same number of games with stationary (S) bases there were ten sliding injuries. To find out how strongly

---

[1]Antleman, G. (1996), *Elementary Bayesian Statistics*, Cheltenham: Edward Elgar Publishing.
[2]Berry, D. (1996), *Statistics: A Bayesian Perspective*, Belmont, CA.: Duxbury Press.

this information indicates that B bases are safer, consider only the twelve injuries. Let p denote the probability that a particular injury results from a breakaway base. Consider the following two prior distributions for p:

p	.1	.2	.3	.4	.5
Prior 1	.1	.1	.1	.1	.6
Prior 2	.2	.3	.2	.1	.1

The person who uses prior 1 believes that there is a high probability that the breakaway bases are not safer than stationary bases but allows for the possibility that they are safer. The person with prior 2 believes apriori that the breakaway bases will help in reducing injuries and therefore places significant probability on small p values. Use the program 'mod_crit' to compare the posterior analyses using the two priors. Do the posterior probabilities depend significantly on the choice of prior? If so, which prior is more consistent with the observed data?

Chapter 10

Learning about Continuous Models

10.1 Introduction

> Minitab command to learn about continuous models:
> **exec 'mod_cont'**

The program 'mod_disc' illustrates learning about one model or parameter when a discrete set of models is used. The program 'mod_cont' implements Bayes' rule when the model values are continuous-valued. Suppose that one chooses a continuous prior density for the unknown model. We assume that one is able to take a simulated sample from this prior density. Minitab has the capability to simulate numbers from a wide variety of distributions, so producing this *prior sample* is usually easy to do. The program 'mod_cont' takes this prior sample and the definition of the likelihood and the data and returns an approximate simulated sample from the posterior distribution. This general method of computation was earlier used in Chapter 5 in the program 'pp_exch' for updating an exchangeable prior for two proportions.

To illustrate how this program works, consider the problem of estimating a proportion that was considered in Chapter 4. Suppose you are interested in estimating the proportion p of times in the long run that a particular coin, when tossed, will result in a head. You think that coins are generally fair, but since this coin is very badly weathered, the proportion p could be anywhere from .4 to .6 with values of .5 most likely. After some thought, you decide that a beta (10, 10) density is a reasonable approximation to your opinion about this proportion.

To explain how probabilities are updated using this method, we will take a tiny simulated sample of ten from this prior density. (In typical problems, a sample size of least 500 should be used.) First define a column 'prior_s' and use the Minitab "rand" command to simulated ten values from a beta (10, 10) curve. The simulated values of p are displayed using a dotplot.

```
MTB > rand 10 'prior_s';
SUBC> beta 10 10.
MTB > dotplot 'prior_s'
```

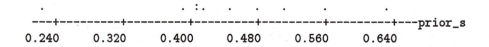

Now suppose you get some data — the coin is tossed twenty times and five heads and fifteen tails are observed. The likelihood, for a given value of the proportion p, is given by $LIKE = p^5(1 - p)^{15}$. For each simulated proportion value, you compute the value of the likelihood. One converts the values of the likelihoods to probabilities by dividing each value by the sum.

```
MTB > let 'like'='prior_s'**5*(1-'prior_s')**15
MTB > let 'like'='like'/sum('like')
```

This table shows the simulated values of p in order and the corresponding likelihoods:

p	.23	.40	.41	.41	.41	.45	.48	.51	.56	.64
$LIKE$	.36	.14	.13	.12	.12	.07	.04	.02	.00	.00

These values of the likelihood are informative — the data are most supportive of the proportion value of .23 and values larger than .5 are not likely.

To obtain a simulated sample from the posterior density, one uses a weighted bootstrap method. (This is also called the SIR algorithm.) One takes a sample of size 10 with replacement from the prior sample above, where the probability of choosing a particular value from the prior sample is proportional to its likelihood value. So the chance that the value $p = .23$ will be sampled is 36%, the chance $p = .40$ is sampled is 14%, and so on. This new sample is distributed (approximately) from the posterior density. In Minitab, this sampling is performed using the 'discrete' command; the prior and posterior samples are displayed below. Note that the proportion value of .23 was selected four times and no proportion values larger than .5 were chosen.

```
MTB > Random 10 'post_s';
MTB > discrete 'like'.
MTB > dotplot 'prior_s' 'post_s';
SUBC> same.
```

```
             .                . :.     .     .      .        .          .
          ---+---------+---------+---------+---------+---------+---prior_s
             .
             :
             :                   . :.    .
          ---+---------+---------+---------+---------+---------+---post_s
          0.240     0.320     0.400     0.480     0.560     0.640
```

The program 'mod_cont' implements this weighted bootstrap algorithm. To set up this program, one uses the Minitab rand command to place a simulated sample from the prior density of the model into the column 'prior_s'. Then the program can be executed. The inputs to the program are the same as the discrete modeling program 'mod_disc'. One inputs the number of the likelihood function (there are seven options, including binomial, normal, Poisson, etc.) and the data values. We recommend that one simulate at least 500 values from the prior density — the program will automatically simulate a posterior sample of the same size as the prior sample. The program will summarize the prior and posterior samples using dotplots and the Minitab "describe" command. The simulated values from the posterior are placed in the column 'post_s'.

One attractive feature of this algorithm is that it can be applied for any proper prior distribution and a wide class of likelihood functions. We will illustrate this flexibility in modeling in the examples to follow. Another nice feature is that the posterior distribution is represented by means of a sample of values. One can learn about the posterior distribution in much the same way as one would learn about a sample of data values. Minitab has many tools for learning about a batch of data, and these same tools can be used to explore the posterior simulated sample.

10.2 Learning using capture/recapture sampling

As a first illustration of the use of the 'mod_cont' command, we consider a finite population inference problem similar to that described in Section 9.4. The difference here is that the total size N of the population is unknown, and we learn about the size by means of a capture/recapture sampling scheme. Strictly speaking, the values of N are discrete and the program 'mod_disc' described in Chapter 9 could be used. However, since there are so many possible values for N, one can regard this unknown number as approximately continuous, and it will be convenient to use a continuous prior density to model one's prior beliefs about N.

To illustrate this situation, suppose that you are interested in estimating the total number of fish N in a small lake. To learn about N, you introduce a known number M of marked fish into the lake. You allow the marked fish to swim throughout the lake for some time. Then you take a sample of n from the lake without replacement and count the number x of marked fish in the sample. Suppose that 100 tagged fish are added to the lake, a sample of size 40 is selected, and five tagged fish are counted. Based on this observed value of $x = 5$, what have you learned about the total number of unmarked fish N?

Although the exact value of N is unknown, it is possible that you have some prior information and can make an educated guess about plausible values of N. Suppose that you can model your beliefs about the logarithm of N with a normal prior density with mean m and standard deviation s. To choose this prior density, the program 'normal_s' may be helpful. This program will give the particular normal density that matches two percentiles of the distribution that are input by the user.

Suppose that you guess that the number of fish in the lake N is 1000, and, in addition, you believe, with probability .9, that the number is within 50% of this guess. In other words, you think that the median of the distribution of $\log(N)$ is $\log(1000) = 6.9$ and $\log(N)$ falls in the interval $(\log(500), \log(2000)) = (6.2, 7.6)$ with probability .9. Using the program 'normal_s', you match this information about $\log(N)$ to a normal density curve with $m = 6.9$ and $s = .4$.

First, one must produce a simulated sample from the prior distribution for N. (See the output below.) To do this, you use the Minitab rand command to simulate a sample of size 1000 from the normal prior distribution of $\log(N)$. Then one exponentiates the values to obtain a sample from the prior distribution of N. This prior sample is located in the Minitab column 'prior_s'. In the 'mod_disc' command, one indicates by typing "6" that a capture/recapture likelihood will be used, and the data vector consists of the number of tagged fish (100), the sample size (40), and the number of marked items in the sample (5).

```
MTB > name c1 'prior_s'
MTB > rand 1000 'prior_s';
SUBC> normal 6.9 .4.
MTB > let 'prior_s'=exp('prior_s')
MTB > exec 'mod_cont'

INPUT THE NUMBER OF THE LIKELIHOOD:
(1-Binomial P, 2-Normal M, 3-Poisson L, 4-Hypergeometric S,
 5-Discrete Uniform N, 6-Capture/Recapture N, 7-Exponential M)
DATA> 6

INPUT
   (number of marked items, sample size, number marked in sample)
DATA> 100 40 5

Input number of simulated values:
DATA> 1000

Each dot represents 5 points
                    .:: .
                   :::: :  .
                   ::::::::.:
                 ::::::::::::.:
                 ::::::::::::::
                .:::::::::::::: .
              .:::::::::::::::.:: ..
          ..:::::::::::::::::::::::::.::.....  ......  ..         .
        +---------+---------+---------+---------+---------+-------prior_s
        0       700      1400      2100      2800      3500
```

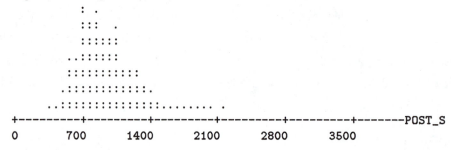

```
Each dot represents 9 points
                        :  .
                       ::: .
                       ::::::
                     ..::::::
                     :::::::::::::
                    .:::::::::::::.
                  ..:::::::::::::::::.........  .
        +---------+---------+---------+---------+---------+-------POST_S
        0        700       1400      2100      2800      3500
```

MTB > describe 'prior_s' 'post_s'

Variable	N	Mean	Median	TrMean	StDev	SEMean
prior_s	1000	1098.7	1027.8	1064.7	455.5	14.4
POST_S	1000	933.51	889.37	918.06	281.45	8.90

Variable	Min	Max	Q1	Q3
prior_s	240.3	3653.6	783.7	1323.9
POST_S	343.47	2194.91	727.47	1089.45

This command places 1000 simulated values of N from the posterior distribution in the column 'POST_S'. The parallel dotplots shown in the output are useful in contrasting graphically the prior and posterior distributions for N. The describe command, which gives descriptive statistics, is helpful in summarizing these two probability distributions. Note that the posterior distribution for N is shifted left from the prior distribution. Initially, you believed that the median for N was approximately 1000 with a 50% chance of falling in the interval (800, 1300). Now, after observing data, you think that N is about 900 and approximately has a 50% chance of falling in the interval (700, 1100).

10.3   Estimating a proportion using a mixture prior

This algorithm is useful in implementing Bayes' rule in situations where a nonconjugate prior is used for the model of interest. Let us return to the baseball example of Chapter 4 where one wishes to learn about the batting average p of a prospective major league player. Suppose, as before, you are pretty sure that the batting average is between .210 and .390. However, you have little knowledge about the batting average if it is under .210 or over .390. We have to do some imagining, since most baseball fans would say that batting averages over, say .500, are impossible. However, let's pretend that you wish to give all batting averages over .390 equal prior probabilities and all batting averages under .210 equal probabilities. One can approximate these prior beliefs by use of a distribution that is a mixture of two beta distributions. Specifically, suppose that the prior distribution is 90% a beta(20.4, 47.6) distribution and 10% a uniform distribution (a beta(1, 1) distribution). A sample from a mixture of two beta distributions is easy to simulate. In the output here, we obtain this sample by combining a sample of 900 values from the beta distribution with a sample of 100 values from a uniform distribution. This simulated sample is placed in the column 'prior_s'.

```
MTB > rand 900 'prior_s';
SUBC> beta 20.4 47.6.
MTB > rand 100 c2;
SUBC> unif 0 1.
MTB > stack 'prior_s' c2 'prior_s'
MTB > exec 'mod_cont'

INPUT THE NUMBER OF THE LIKELIHOOD:
(1-Binomial P, 2-Normal M, 3-Poisson L, 4-Hypergeometric S,
 5-Discrete Uniform N, 6-Capture/Recapture N, 7-Exponential M)
DATA> 1

INPUT
  (number of successes, number of failures)
DATA> 30 20

Input number of simulated values:
DATA> 1000

Each dot represents 8 points
                .:: .
                :::: .
               : : : : : :
               : : : : : : .
              .: : : : : : :
              : : : : : : : : .
            : : : : : : : : : :
     ... ....:::::::::::::::.... . .... ............ ....
     +---------+---------+---------+---------+---------+-------prior_s
    0.00      0.20      0.40      0.60      0.80      1.00
```

Each dot represents 14 points

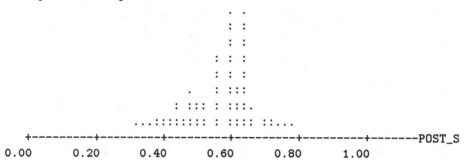

```
                              . .
                            : :   :
                            : :   :
                          : : :   :
                          : : :   :
                  .       : : :::
              : ::: :  :::.
          ...:::::::::: : :::: ::...
      +---------+---------+---------+---------+---------+-------POST_S
    0.00      0.20      0.40      0.60      0.80      1.00
```

Suppose that the baseball player is unusually hot during an observed fifty at-bats and gets thirty hits, for an observed batting average of .600. To obtain the posterior distribution, one runs the program 'mod_cont' — one types 1 to choose a binomial likelihood and enters the number of successes and failures (thirty hits and twenty outs) on the DATA line. As in the previous example, the simulated samples from the prior and posterior distribution are displayed using dotplots. The interesting aspect here is how the prior information and data are combined in the posterior distribution. Note that the posterior distribution is centered about the value .6 — the information in the data is dominant, and the prior information that p likely lies between .210 and .390 is ignored. A very unusual datum outcome has been observed, and since this indicates a value of p in a region where one has little prior knowledge, the posterior distribution is approximately the same as the likelihood function.

10.4   Exercises

1. (O'Hagen[1], exercise 9.d.1) A sociologist is studying recidivism. She has a sample of fifteen people who were convicted of arson and released from prison twenty years previously. Assume that the number of convictions for these fifteen people comes from a Poisson model with parameter L. Suppose that a gamma distribution with mean 1.5 and variance 2.0 reflects the sociologist's prior beliefs about L. The observed data are

$$0, 1, 0, 2, 3, 0, 0, 1, 2, 2, 0, 1, 2, 1, 1$$

---

[1] O'Hagan, A., *Probability: Methods and Measurement*, New York: Chapman and Hall.

(a) Find the posterior distribution for L by use of macro 'mod_cont'. Simulate 1000 values from the gamma prior distribution and use the program 'mod_cont' with a Poisson likelihood.

(b) Summarize the simulated sample from the posterior distribution to learn about L. In particular, find an estimate for L and an interval for L that contains at least 90% of the posterior probability.

2. (Estimating a truncated normal mean) (Berry[2], exercise 11.2) A study evaluated a drug called amiloride in patients who were suffering from cystic fibrosis. Measurements were made of the patients' forced vital capacities after six months of using an aerosol spray containing amiloride and after six months of using a spray without amiloride (vehicle only). The numbers below are the differences (amiloride − vehicle) for the fourteen patients:

$$-165, 300, 550, -45, -124, 33, -155, -228, 86, 123, 70, 262, 243, 46$$

Let M denote the mean difference in forced vital capacity for the population of cystic fibrosis patients. We assume that the observed measurements are normally distributed with mean M and known standard deviation 150. Suppose that a doctor believes (with probability 1) that the drug is beneficial ($M > 0$); moreover, he assigns M a normal prior distribution with mean 50 and scale parameter 50 that is truncated below by 0. Find the posterior distribution of M and summarize the posterior distribution with the mean and standard deviation. (To use the program 'mod_cont', first simulate 1000 values from a normal distribution with mean 50 and standard deviation 50. Remove the simulated values that are negative; the remaining values are a sample from the prior distribution. Then run the program 'mod_cont' with a normal likelihood and the data consisting of the observed sample mean, the sample size, and the known population standard deviation.)

3. (Estimating a normal mean using a Cauchy prior) Suppose that you observe one observation x from a normal density with mean M and standard deviation 1. Consider the following two prior distributions for M: (1) Cauchy with median 0

---

[2]Berry, D. (1996), *Statistics: A Bayesian Perspective*, Belmont, CA.: Duxbury Press.

and scale parameter 1; (2) normal with mean 0 and standard deviation 1.49. Both prior distributions match the belief that the median of M is 0 and the quartiles are -1 and 1.

(a) Suppose you observe $x = 0$. Using the 'mod_cont' program, compare the posterior distributions using the Cauchy and normal priors. Summarize each of the two simulated posterior samples by a mean and standard deviation.

(b) Repeat part (a) for the case where you observe $x = 5$.

(c) Based on these calculations, comment on the difference between the use of a normal prior and a Cauchy prior.

4. (Estimating the number of successes in a finite population using an informative prior) Consider again the example in Section 9.4 where one wishes to learn about the number of voters S in favor of a school levy in a small community of 100 voters. Suppose that one has some prior knowledge about S. Specifically, suppose that one's belief in the proportion $p = S/100$ of voters in favor is modeled by a beta(5, 5) distribution. As in the example, suppose you take a sample of twenty voters and twelve are in favor. Find the posterior distribution for S using this informative prior. (First simulate 1000 values from the beta prior for p. If this simulated sample is stored in the column 'P', a sample of values from the prior distribution of S in the column 'prior_s' is obtained using the Minitab command

```
let 'prior_s'=round(100*'p')
```

Finally, run 'mod_cont' using the prior sample in 'prior_s' and a finite population likelihood.)

5. (Antleman[3], Chapter 8) A seller receives 800-number telephone orders from a first geographic area at a rate of L_1 per week and from a second geographic area at a rate of L_2 per week. Assume that incoming orders behave as if generated by a Poisson distribution and that, in the past, the ordering rate in area one has been about 50% higher than the ordering rate in the second area. Suppose a series

---

[3]Antleman, G. (1996), *Elementary Bayesian Statistics*, Cheltenham: Edward Elgar Publishing.

of newspaper ads are run in the first area for a period of four weeks, and sales for these four weeks are 260 units in area 1 and 165 units in area 2. The seller is interested in the effectiveness of these ads. One measure of this would be the probability that the sales rate in area 1 is greater than 1.5 times the sales rate in area 2:

$$P(L_1 > 1.5L_2)$$

Before the ads run, the seller has assessed a prior distribution for L_1 to be gamma with parameters 144 and .417, and the prior for L_2 to be gamma (100, .4).

(a) Since the posterior distributions for L_1 and L_2 can be assumed independent in this case, the joint posterior for (L_1, L_2) is determined by the marginal posterior densities for the two parameters. To obtain a simulated sample of L_1 values, place a large number of values of a gamma $(144, 417)$ density into the column 'PRIOR_S' and run 'mod_cont'. Copy the posterior simulated sample in 'POST_S' to the column 'L1' using the command

```
copy 'POST_S' 'L1'
```

In a similar fashion, take a sample of values from the posterior density of L_2 — copy the values in the column 'L2'. (The size of this sample should be the same as the one taken from the distribution of L_1.)

(b) The probability of interest can be computed and displayed using the commands

```
let k1=mean('L1'>1.5*'L2')
print k1
```

Chapter 11

Summarizing Posterior Distributions

11.1 Introduction

The Minitab programs in this chapter can be used to summarize an arbitrary one- or two-parameter continuous posterior distribution. Suppose that the parameter of interest is denoted by M. One assumes that M is continuous-valued, and prior information about M is represented by means of a density $g(M)$. To obtain more information about the parameter, some data are collected. The likelihood $l(M)$ is the probability of observing this data for a given value of the parameter M. Then, by Bayes' theorem, the updated density for M is proportional to the product of the prior density and the likelihood:

$$h(M|\text{data}) \propto g(M)l(M).$$

The posterior density can be written in the equivalent form

$$h(M|\text{data}) = \frac{g(M)l(M)}{H},$$

where H is the unknown normalizing constant

$$H = \int g(M)l(M)dM.$$

All of the information about the parameter M is contained in the posterior probability density h. This distribution is used to find a "best guess" for the parameter and to find an interval estimate that contains the parameter with a given probability. The distribution can also be used to make decisions. One can test the hypothesis that M is smaller than a particular value, say M_0, by computing the posterior probability of this set.

In many cases, the likelihood and prior will be of convenient functional forms, and one can learn about the posterior density by analytical methods. For example, if a beta density is used to represent prior opinion about a proportion p, and a binomial sample is taken, then the posterior density will also be a beta density. This is a familiar density that is easy to summarize. In other cases, however, choices for the likelihood and prior will be made where the posterior density will not have a nice functional form. In these cases, the programs of this chapter are useful for numerically summarizing this density function.

One uses these programs by defining the posterior density function in a Minitab macro file. Details of how one defines the density function and how it is placed in a file are described in Section 11.2. Then the Minitab programs are designed to perform different computational algorithms for the posterior density that is defined in this file. The programs 'laplace1' and 'laplace2' implement the Laplace algorithm for a one- and two-parameter posterior density, respectively. The programs 'ad_quad1' and 'ad_quad2' illustrate the use of an adaptive quadrature method, and the programs 'metrop' and 'gibbs' use general-purpose simulation algorithms.

The programs are illustrated in this chapter for two examples. For a one-parameter example, suppose that one is interested in estimating the probability p that a particular coin lands heads. Suppose that the prior beliefs of the user are stated in terms of the logit of the probability $T = \log(p/(1-p))$. Before taking any data, the person assigns to the logit parameter T a Cauchy prior with median 0:

$$g(T) = (1 + T^2)^{-1}.$$

This density reflects the belief that the logit of the probability is close to 0. This is equivalent to believing that the probability p is close to .5. Next, suppose that the coin is tossed ten times, and one head and nine tails are observed. Since the user likes to think in terms of logits, he is interested in computing the posterior mean and standard deviation of the logit parameter T. Also, he is interested in seeing whether the coin is significantly biased. Suppose that he thinks the coin is biased if the logit T is outside of the interval $(-.5, .5)$. He would like to compute the posterior probability that T is in the interval $(-.5, .5)$. If this probability is unusually small, then he has reason to believe that the coin is not fair.

To illustrate a two-parameter problem, consider the problem of estimating the association structure in a 2×2 contingency table. Suppose that two independent binomial samples are taken with respective probabilities p_1 and p_2. A common measure of association in a 2×2 table is the log odds-ratio, which is defined as

$$W = \log\left(\frac{p_1/(1 - p_1)}{p_2/(1 - p_2)}\right).$$

Suppose that initially one has little prior information about the location of either proportion. One standard choice of a noninformative prior density for a proportion is the limit of a beta(a, b) density as the parameters a and b approach 0. If the two probabilities are assumed to be independent with each distributed according to this noninformative density, one obtains the improper joint density:

$$g(p_1, p_2) = \frac{1}{p_1(1 - p_1)p_2(1 - p_2)}.$$

Suppose that one observes one success and ten failures in the first sample and three successes and twelve failures in the second sample. One would like to learn about the association in the table by the computation of the marginal posterior density for the log odds-ratio W. Specifically, one may be interested in summarizing this density for W using the posterior mean and posterior standard deviation. If there is independence in the table, the log odds-ratio W is equal to 0. Since there is some evidence that there is a negative association structure in the table, one is interested in computing the posterior probability that W is smaller than 0.

11.2 Setting up the problem

To use the programs of this chapter for a particular problem, one must initially place the definition of the posterior density into a Minitab macro. The definition of a one-parameter posterior is located in the macro 'logpost1' and the definition for a two-parameter problem is contained in the macro 'logpost2'.

All of the computational methods described in this chapter assume that each parameter is real-valued. Thus it may be necessary to reexpress the parameters into real-valued ones by use of an appropriate transformation. For example, if the parameter is a proportion p that is restricted between 0 and 1, one can transform it by the logit function $\log(p/(1-p))$ to make it real-valued. A standard deviation S that takes only positive values can be transformed to a real-valued parameter by the logarithm $\log(S)$. In most applications, a suitable transformation can be found so that the resulting reexpressed parameters can take on all real values. This assumption that the parameters are real-valued is made primarily for computational reasons. The methods generally give more accurate approximations to the posterior density when the parameters are real-valued. Also, this restriction does not mean that one can only learn about the transformed parameters — the methods will give accurate approximations to the posterior density of the original parameters of interest.

These methods also assume that the definition of the natural logarithm of the posterior density is placed in the macros. This is convenient again for computational reasons. Values of the log posterior density are generally more stable and therefore less apt to underflow or overflow than values of the density.

In the first example, one is interested in learning about a proportion p. The prior density for logit T is assumed to be Cauchy with median 0 and scale 1, and one observes an independent sequence of one success and nine failures. The likelihood function as a function of p is given by

$$l(p) = p^1(1-p)^9.$$

Equivalently, since $p = \exp(T)/(1+\exp(T))$, the likelihood as a function of the logit parameter T is given by

$$l(T) = \frac{\exp(T)}{(1+\exp(T))^{10}}.$$

Since the logit T is real-valued, we express the posterior density in terms of T. The posterior density of T is proportional to the product of the likelihood $l(T)$ and the Cauchy prior density $g(T) = \frac{1}{1+T^2}$. The logarithm of the posterior density, therefore, is given (up to an additive constant) by

$$\log h(T|\text{data}) = T - 10\log(1+\exp(T)) - \log(1+T^2).$$

The definition of the logarithm of the posterior density for a one-parameter problem is contained in the file named 'logpost1.MTB'. This file for this problem is listed here. The input to the macro is a Minitab column 'x' of values of T and the output is a column 'f' of the corresponding values of the posterior density. (These particular column names 'x' and 'f' should always be used for the parameter value and log posterior values.) The numbers of observed successes and failures are stored in the Minitab constants k11 and k12, respectively. So this program is easy to modify for other data values. The main part of the macro contains three "let" commands. The first two commands define the values of the two Minitab constants and the last let command computes the column of values of the log posterior.

```
MTB > type 'logpost1.MTB'
let k11=1
let k12=9
let 'f'=k11*'x'-(k11+k12)*log(1+exp('x'))-log(1+'x'**2)
```

A similar procedure is used to set up a two-parameter problem. In the example of the introduction, one is interested in looking at the association structure in a 2×2 table. The two parameters of the problem are the two binomial probabilities p_1 and p_2. The number of successes and failures in the two samples are $(1, 10)$ and $(3, 12)$. The likelihood function of p_1 and p_2 is given by

$$l(p_1, p_2) = p_1^1(1 - p_1)^{10}p_2^3(1 - p_2)^{12},$$

and the prior for (p_1, p_2) is given by the noninformative form $g(p_1, p_2) = \frac{1}{p_1(1-p_1)p_2(1-p_2)}$. The posterior density for the two probabilities is given by the product

$$h(p_1, p_2|\text{data}) = g(p_1, p_2)l(p_1, p_2) = p_1^0(1 - p_1)^9 p_2^2(1 - p_2)^{11}.$$

To convert to real-valued parameters, we transform the probabilities to the logits $T_1 = \log(p_1/(1 - p_1))$ and $T_2 = \log(p_2/(1 - p_2))$. Last, since we are interested in learning about the log odds-ratio $W = \log(\frac{p_1/(1-p_1)}{p_2/(1-p_2)}) = T_1 - T_2$, we transform (T_1, T_2) into the difference of logits W and the sum of logits $U = T_1 + T_2$. The resulting posterior density of (W, U) is given by

$$h(W, U|\text{data}) = \exp(T_1)(1 + \exp(T_1))^{-11} \exp(3T_2)(1 + \exp(T_2))^{-15},$$

where $T_1 = (W + U)/2$ and $T_2 = (U - W)/2$. The logarithm of the posterior is given by

$$\log h(W, U|\text{data}) = T_1 - 11\log(1 + \exp(T_1)) + 3T_2 - 15\log(1 + \exp(T_2)).$$

In the following output, the definition of the posterior for this example is contained in the file named 'logpost2.MTB', whose contents are listed. The input to the macro is two columns — values of the first parameter (W) in column 'x' and values of the second parameter (U) in column 'y'. The data values are placed in the column 'data' and the individual numbers of successes and failures for the two samples are stored in the Minitab constants k21, k22, k23, k24. In the macro, let commands are used to store values of T_1 and T_2 in columns c201 and c202. A final let command places the values of the log posterior in the column 'f'.

```
MTB > type 'logpost2.MTB'
set 'data'
1 10 3 12
let k21='data'(1)
let k22='data'(2)
let k23='data'(3)
let k24='data'(4)
let c201=('x'+'y')/2
let c202=('y'-'x')/2
let 'f'=k21*c201-(k21+k22)*log(1+exp(c201))+k23*c202-(k23+k24)*log(1+exp(c202))
```

11.3   The Laplace method

> Minitab programs to implement the Laplace method:
> For a one-parameter problem:
> **exec 'laplace1'**
> For a two-parameter problem:
> **exec 'laplace2'**

The programs 'laplace1' and 'laplace2' use a Newton-Raphson algorithm to search for the mode or most likely value of the posterior density. In the one-parameter case, let M denote the parameter, and let M_0 represent a current guess at the value of the mode. Then, using the Newton-Raphson algorithm, the next guess at the mode is given by

$$M_1 = M_0 - \frac{f_1(M_0)}{f_2(M_0)},$$

where $f_1(M_0)$ and $f_2(M_0)$ are the first and second derivatives of the logarithm of the posterior density evaluated at the current guess M_0. Starting with an initial guess at the mode, the algorithm is iterated a number of times. At each step of the algorithm, the program 'laplace1' gives the current estimate at the posterior standard deviation. For the two-parameter problem, the program 'laplace2' outputs the two standard deviations and covariance between the parameters. Both programs also give the Laplace estimate of the logarithm of the normalizing constant H of the posterior density.

The Laplace method provides a simple normal curve summary of the posterior density. In the one-parameter case, if m is the mode, then one can approximate the posterior density by a normal density with mean m and standard deviation $s = \sqrt{-1/f_2(m)}$. This summary can be used for a variety of posterior inferences. For example, in our one-parameter binomial example, one can assume that the logit T is approximately normally distributed with parameters m and s. One can learn about the proportion of interest p by making the suitable transformation of a normal distribution on T. A simple way of doing this on Minitab is by simulation. One simulates a large number of values from the approximate normal posterior distribution for T. Transform this column of T to p values using the transformation $p = \exp(T)/(1 + \exp(T))$. The values of p obtained are (approximately) a simulated sample from the posterior distribution.

The program 'laplace1' will perform Newton-Raphson steps on a one-parameter posterior density that is defined in the macro 'logpost1'. There are two inputs to this program — a guess at the posterior mode and the number of Newton-Raphson steps to perform.

A run of 'laplace1' for the binomial example is shown in the following output. One starts with a guess of 0 at the mode and the program performs five Newton-Raphson steps. The succeeding guesses at the mode appear to settle down after only a few iterations. The posterior mode is -1.42, and the associated estimate at the standard deviation is .87. The estimate at the logarithm of the normalizing constant is -3.91. Recall that one was interested in computing the posterior probability $P(|T| < .5)$ that the coin was roughly fair. Assume that the posterior distribution for T is normally distributed with mean -1.42 and standard deviation .87. Then, using normal tables (or the program 'normal' described in Chapter 6), the probability of interest is computed to be .128.

```
MTB > exec 'laplace1'

INPUT GUESS AT POSTERIOR MODE:
DATA> 0

INPUT NUMBER OF ITERATIONS:
DATA> 5

   Row       MODE        STD  LOG_INTG
     1  -0.886723   0.470836  -6.76578


   Row       MODE        STD  LOG_INTG
     1   -1.30571   0.673015    -4.394


   Row       MODE        STD  LOG_INTG
     1   -1.41766   0.819856  -3.97834


   Row       MODE        STD  LOG_INTG
     1   -1.42373   0.857813  -3.92344


   Row       MODE        STD  LOG_INTG
     1   -1.42381   0.868545  -3.91098
```

The program 'laplace2' implements the Newton-Raphson algorithm for a problem with two parameters where the log posterior density is stored in the macro 'logpost2'. The input to 'laplace2' is two numbers, the guess at the mode of the two-parameters. As in the previous command, one inputs the number of times the procedure is iterated.

For the contingency table example, one guesses initially that the mode is $(W, U) = (1, 1)$ and runs the algorithm for a total of five iterations (see the output below). The guesses at the posterior means and standard deviations appear to converge after three iterations. The output of this command is the mode of (W, U), the estimated standard deviations and covariance between the two parameters, and the Laplace estimate at the normalizing constant of the density.   Since we wish to estimate the association in the table, we note that the posterior mode of the log odds-ratio W is $-.92$ with an associated standard deviation of 1.24. The Laplace approximation is that the log odds-ratio is normally distributed with this mean and standard deviation. To see whether the two variables are negatively associated, one computes $P(W < 0)$ using the program 'normal'. From the output we see that this probability is .77 — this is probably not large enough to conclude that the association in the table is significant.

```
MTB > exec 'laplace2'

INPUT GUESS AT MODE:
DATA> 1 1

INPUT NUMBER OF ITERATIONS:
DATA> 5

    Row      MN_1     STD_1      MN_2     STD_2      COVAR  LOG_INTG
      1  -1.04684   0.85129  -3.43018   0.85129    0.19511  -22.3648


    Row       MN_1     STD_1      MN_2     STD_2      COVAR  LOG_INTG
      1  -0.927094   1.20075  -3.67809   1.20075    0.69814  -8.83597


    Row       MN_1     STD_1      MN_2     STD_2      COVAR  LOG_INTG
      1  -0.916306   1.23626  -3.68887   1.23626   0.702681  -8.71386


    Row       MN_1     STD_1      MN_2     STD_2      COVAR  LOG_INTG
      1  -0.916094   1.24168  -3.68866   1.24168   0.689256  -8.69772


    Row      MN_1     STD_1      MN_2     STD_2      COVAR  LOG_INTG
      1  -0.91617   1.28012  -3.68882   1.28012   0.799844  -8.66131

MTB > exec 'normal'

INPUT THE VALUES OF THE MEAN AND STANDARD
DEVIATION OF THE NORMAL DISTRIBUTION:
DATA> -.92 1.24

TYPE 'y' TO COMPUTE CUMULATIVE PROBABILITIES:
----------------------------------------------------------------
   Input values of M of interest.  The output is the column of
   values M and the column of cumulative probabilities PROB_LT.
----------------------------------------------------------------
y
DATA> 0
DATA> end

ROW     M    PROB_LT
  1     0    0.770937
```

11.4   An adaptive quadrature routine

> Minitab command to implement adaptive quadrature:
> For a one-parameter problem:
> **exec 'ad_quad1'**
> For a two-parameter problem:
> **exec ad_quad2'**

Although the Laplace method is useful in finding the mode of the posterior density, it makes the implicit assumption that the density is normally distributed and this assumption may be inaccurate for some problems. The Minitab programs described in this section use a general numerical integration method, adaptive quadrature, to compute the normalizing constant of the posterior density that is defined in 'logpost1' (one-parameter problem) or 'logpost2' (two-parameter problem). This method gives estimates at the normalizing constant, the posterior mean, and standard deviation that generally are more accurate than the estimates found using the Laplace method.

First consider a problem for a single parameter M where the likelihood function is given by $l(M)$ and the prior density by $g(M)$. This method is based on approximating the normalizing constant H of the posterior density by the sum

$$H = \int l(M)g(M)dM \approx \sum_{i=1}^{K} w_i l(M_i)g(M_i),$$

where $M_1, ..., M_K$ are the parameter values where the function is evaluated (the grid), and $w_1, ..., w_K$ are weights defined to get an accurate estimate at the integral.

To use this algorithm, one first inputs guesses at the mean and standard deviation of the posterior density. Based on these guesses, a grid of values for the parameter is constructed and the integral is approximated by the sum given. This grid is used to obtain new estimates at the posterior mean and standard deviation. If one continues this procedure, then these new values of the mean and standard deviation are used to construct a new grid, and this procedure is iterated until it stabilizes.

The command 'ad_quad1' is illustrated for the one-parameter binomial problem (see the output below). There are two inputs: the guess at the posterior mean and standard deviation of the posterior distribution and the number of iterations of the algorithm. Here we use the somewhat naive guesses of 0 and 1 and iterate the procedure five times. The program displays the estimated values of the posterior mean and standard deviation and the estimate at the logarithm of the normalizing constant H at each step of the algorithm. In addition, for each iteration, the posterior density is plotted on the ten-point grid that is constructed. The final plot is shown in Figure 11.1. Note that the posterior estimates settle down quickly. The posterior mean and standard deviation are given by -1.80 and 1.02; these estimates can be compared with the values -1.42 and $.87$ given by the Laplace method. Note from the graph that the posterior density displays some left skewness; this explains why the posterior mean is smaller than the posterior mode. If one assumes that the posterior density is normally distributed with mean -1.80 and standard deviation 1.02, then the probability that T is within $.5$ of 0 is given by $.089$.

The elements of this integration computation are stored in three Minitab columns. The values of the parameter at the grid are given in the column 'X', the values of the density are given in the column 'DENSITY', and the weights used in computing the normalizing constant are stored in the column 'WX'.

```
MTB > exec 'ad_quad1'

INPUT MEAN AND STANDARD DEVIATION:
DATA> 0 1

INPUT NUMBER OF ITERATIONS:
DATA> 5

Row      MEAN        STD    LOG_INTG
  1   -1.78599   0.99106   -3.86880
  2   -1.80105   1.02136   -3.86742
  3   -1.80249   1.02268   -3.86801
  4   -1.80254   1.02275   -3.86802
  5   -1.80254   1.02275   -3.86803
```

This algorithm is implemented for a two-parameter problem using the command 'ad_quad2'. The first inputs are the guesses at the mean and standard deviation for the first parameter, the guesses at the mean and standard deviation for the second parameter, and the covariance between the two parameters. One also indicates the number of iterations of the algorithm. The output of this program are the updated estimates at the posterior moments and the logarithm of the normalizing constant after

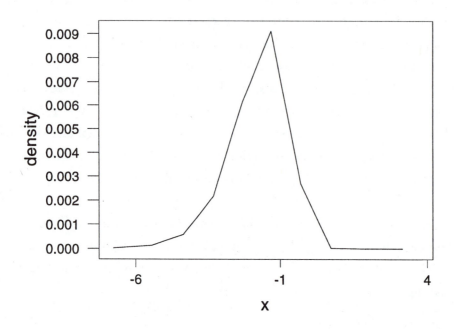

Figure 11.1: Posterior density of logit parameter on ten-point grid for binomial example.

each iteration. In addition, the joint probability density is graphed using a scatterplot. Also the values of the parameters on the grid and the posterior density are stored in Minitab columns. The columns 'X' and 'Y' store the values of the two parameters, and 'F' contains values of the density function.

In the contingency table example, we run the command 'ad_quad2' for five iterations with starting guesses at the moments given by 0, 1, 0, 1, 0 (see the output below). Based on these guesses at the moments, the program sets up a grid of values for the two parameters on which to compute the density. A summation formula similar to that described earlier is used to compute the normalizing constant and new estimates at the posterior moments at the density. These new estimates are used to construct a new grid of values, and the procedure repeats the process at the next iteration. The output of this program is the posterior moment estimates and the estimate of the logarithm of the normalizing constant H at each iteration. For this example, note that the estimates of the posterior means and standard deviations are significantly different than the estimates given in the Laplace method. The estimated mean and standard deviation of the log odds-ratio W using the Laplace method are $-.92$ and 1.24; in contrast, the estimates using adaptive quadrature are -1.32 and 1.51.

One can explain the differences between the Laplace and adaptive quadrature algorithms by looking at the graph of the joint posterior density in Figure 11.2. The dots in the figure represent the values of the two parameters on which the posterior density is computed. Values of the first parameter (W) are displayed along the horizontal axis and values of the second parameter (U) along the vertical. The value of the density with the largest value (approximately the mode) is displayed with a black dot; the next darkest symbols correspond to density values within 10% of the largest density value, and the x's correspond to density values within 1% to 10% of the modal value. This graph clearly shows that W and U are negatively correlated, and both parameters display left skewness. Since the adaptive quadrature method uses more information than the Laplace method about the posterior density, the estimates of the posterior moments and the normalizing constant are expected to be more accurate than the values given by the Laplace method.

```
MTB > exec 'ad_quad2'

INPUT MX, SX, MY, SY, COV:
DATA> 0 1 0 1 0

INPUT NUMBER OF ITERATIONS:
DATA> 5
```

Row	MN_1	STD_1	MN_2	STD_2	COVAR	LOG_INTG
1	-1.12931	1.25898	-4.13558	0.93885	0.36845	-8.44208
2	-1.34772	1.54778	-4.39761	1.55365	1.42565	-8.59427
3	-1.31503	1.50677	-4.35515	1.50729	1.30219	-8.60408
4	-1.31601	1.50882	-4.35637	1.50916	1.30769	-8.60390
5	-1.31596	1.50873	-4.35631	1.50908	1.30744	-8.60390

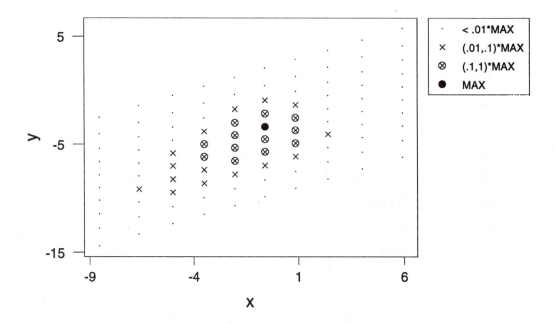

Figure 11.2: Posterior density of two parameters computed on a grid of values for contingency table example.

11.5   Simulation

> Minitab command to implement simulation methods:
> Metropolis algorithm for a one-parameter problem:
> **exec 'metrop'**
> Gibbs sampling for a two-parameter problem:
> **exec 'gibbs'**

A third general method of summarizing posterior distributions is by simulation. Simulation methods have been illustrated for different problems in this book. The Minitab commands 'metrop' and 'gibbs' implement general methods that can be used to simulate from arbitrary one- and two-parameter posterior densities. As before, the particular problem is first set up by defining the posterior density in either the macro 'logpost1' or 'logpost2'.

The program 'metrop' implements the Metropolis-Hastings random walk chain to simulate from a probability distribution of one real-valued parameter M. This simulation algorithm basically "walks through" the probability distribution by use of a random mechanism. Suppose that the algorithm starts at a given value $M^{(0)}$. Then the next candidate from the posterior density is the value

$$M^C = M^{(0)} + cZ,$$

where Z is a standard normal random number and c is a given scale parameter. One computes the probability

$$PROB = \text{minimum of} \{ \frac{g(M^C)l(M^C)}{g(M^{(0)})l(M^{(0)})}, 1 \},$$

and chooses a random uniform number U. If U is smaller then the probability $PROB$, then the algorithm will step to the candidate value M^C; otherwise, the algorithm will remain at the current value $M^{(0)}$. The resulting value of the parameter is called $M^{(1)}$. One continues this random walk for successive iterations. One generates a new candidate value M^C, decides to move to that value or stay at the current value $M^{(1)}$, and $M^{(2)}$ is the location after this decision.

Suppose that one wishes to simulate a sample from the posterior density that is defined in the macro 'logpost1'. One types the command 'exec metrop'. There are three inputs to the program: the size of the simulated sample, the scale factor c for the normal increment density used by the Metropolis-Hastings algorithm, and the starting parameter value $M^{(0)}$ for the simulation. The simulated sample of the given size will be stored in the Minitab column 'POST_S'.

The output from running 'metrop' for the binomial example is shown here. We decide to take a sample of size 1000, the scale parameter c is set equal to 1 (this choice will be suitable for many problems), and we start the random walk at the value $T = 0$.

```
MTB > exec 'metrop'

INPUT NUMBER OF ITERATIONS:
DATA> 500

INPUT SCALE OF NORMAL INCREMENT DENSITY:
DATA> 1
```

```
INPUT STARTING VALUE:
DATA> 0

MTB > dotplot 'POST_S'

Each dot represents 7 points
```

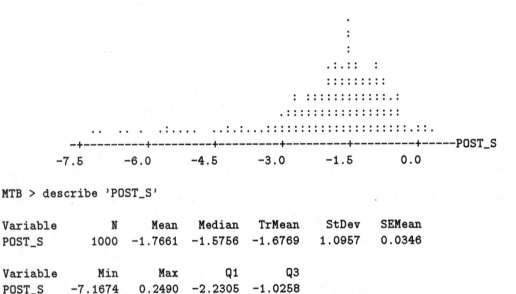

```
MTB > describe 'POST_S'

Variable        N      Mean    Median    TrMean    StDev    SEMean
POST_S       1000   -1.7661   -1.5756   -1.6769   1.0957   0.0346


Variable      Min      Max       Q1        Q3
POST_S    -7.1674   0.2490  -2.2305   -1.0258
```

When a simulated sample is taken from a posterior density, one can summarize the density by performing descriptive and graphical data analysis on the batch of simulated values. In the output here, the Minitab "dotplot" command is used to display the simulated values for the logit parameter T. The "describe" command is used to compute various summary statistics. The general features of the posterior density that were discovered using the adaptive quadrature algorithm are also evident here. The density appears to show some left skewness — this is evident from the general shape of the dotplot and the observation that the posterior mean is smaller than the posterior median. The posterior median is estimated to be -1.58, and the middle 50% of the probability distribution falls in the interval $(-2.23, -1.03)$.

One nice feature of this method is that it is easy to use the output to make inferences about any function of the parameter of interest. In our example, one wishes to learn about the proportion p. In the output below, column C1 is named 'p' and the function p

$= \exp(T)/(1+\exp(T))$ is applied to the simulated values of T that are contained in the column 'POST_S'. The values in 'p' are distributed (approximately) from the posterior distribution of the proportion. Summarizing this column by a dotplot and the describe command, we see that most of the simulated values fall in the interval $(0, .3)$. Exploring further, we see that 83% of the simulated values for p are smaller than .3, so $(0, .3)$ is an approximate 83% probability interval for this proportion.

```
MTB > name c1 'p'
MTB > let 'p'=exp('POST_S')/(1+exp('POST_S'))
MTB > dotplot 'p'
```

Each dot represents 3 points

```
                   .        :    ..
                   :     . : ::
                 : :    : :..::
                 :::..:::::::::
                 ::::::::::::::  ::. :
          : :::::::::::::::::::.:::.:        .
          ::::::::::::::::::::::::::... : .
          :::::::::::::::::::::::::::::.: :..      .   .
          ::::::::::::::::::::::::::::::: .: :. .: .
        +---------+---------+---------+---------+---------+------p
      0.00      0.12      0.24      0.36      0.48      0.60
```

```
MTB > describe 'p'
```

Variable	N	Mean	Median	TrMean	StDev	SEMean
p	1000	0.18555	0.17141	0.17970	0.11606	0.00367

Variable	Min	Max	Q1	Q3
p	0.00077	0.56192	0.09704	0.26389

The program 'gibbs' can be used to obtain a simulated sample from an arbitrary two-parameter posterior density. One simulates values from the posterior density by means of a method called Gibbs sampling. Similar to the Metropolis-Hastings algorithm, the Gibbs sampler is a special type of random walk through a two-parameter probability distribution. The method simulates in turn from two conditional probability distributions. To briefly describe this procedure, consider our example, where the two real-valued parameters are U and W. Suppose that we start the sampling procedure at the values $U^{(0)}$ and $W^{(0)}$. One first simulates a value from the posterior density of the first parameter U conditional on the current value of the second parameter $W^{(0)}$. (This one-parameter simulation is performed using the Metropolis-Hastings algorithm described earlier.) Let the simulated value of U be denoted by $U^{(1)}$. Then one simulates next from the density of the second parameter W conditional on the current value of the first parameter $U^{(1)}$ — call this simulated value $W^{(1)}$. The simulated pair $(U^{(1)}, W^{(1)})$ represents the result of one cycle of the Gibbs sampler. One continues this procedure for many cycles until a large number of simulated pairs $(U^{(1)}, W^{(1)}), (U^{(2)}, W^{(2)}), \ldots$ has been collected.

To use this program, one first places the definition of the logarithm of the two-parameter posterior density in the macro 'logpost2'. Then one runs the program 'gibbs'. One first inputs the number of cycles (iterations) of the Gibbs sampler. Next, one inputs the scale factors for the Metropolis-Hastings routines to simulate from the two one-parameter conditional distributions. For many problems, the choice of scale factors each equal to 1 will be suitable. Last, one inputs a starting location for the two parameters. The choice of starting location often has little impact on the final set of simulated values, but one should in practice rerun the simulation with different starting locations. When the program is completed, the simulated sample of the two parameters is stored in the columns 'POST_X' and 'POST_Y'.

The following output displays the run of 'gibbs' for the contingency table example. In the input, we indicate that we wish to take a simulated sample size of 1000, the scale factors are 1 and 1, and we start the simulation at $(U, W) = (0, 0)$. Figure 11.3 displays a scatterplot of the simulated values of U and V. (The column 'POST_X' contains simulated values of the posterior density of the first parameter, U, and 'POST_Y' corresponds to V.) If one compares the scatterplot with the graph of the posterior density using the 'ad_quad2' command, one sees the same general pattern in both plots. The two parameters are negatively correlated with each parameter displaying skewness toward smaller values. The describe and dotplot commands are used to summarize the simulated samples for U and V.

```
MTB > exec 'gibbs'

INPUT NUMBER OF ITERATIONS:
DATA> 1000
```

```
INPUT SCALES OF NORMAL INCREMENT DENSITIES:
DATA> 1 1

INPUT STARTING VALUE:
DATA> 0 0

MTB > plot 'post_y'*'post_x'

MTB > dotplot 'POST_X' 'POST_Y'

Each dot represents 4 points

                                                 .. .
                                       :    .::.:.
                                      : : :::::::::
                                 . :. : : :::::::::. .
                                 ::::::::::.:::::::::::::::.:.
                                 ::::::::::::::::::::::::::::
                                 .::::::::::::::::::::::::::::.
                            .. ....:.:.:::::::::::::::::::::::::::::::.... .
               +---------+---------+---------+---------+---------+-------post_x
             -8.0      -6.0      -4.0      -2.0       0.0       2.0
Each dot represents 4 points

                                      . ::   .
                                 .   : ::: :
                                 :::::::::::. :  :
                               : :::::::::::.:  :
                             . :::::::::::::::::..:
                            .:.:::::::::::::::::::::::
                       .    .:::::::::::::::::::::::::::.
                 . .:..:::.:::::::::::::::::::::::::::::::.:.. .. .
               +---------+---------+---------+---------+---------+-------post_y
             -10.0      -8.0      -6.0      -4.0      -2.0       0.0

MTB > describe 'POST_X' 'POST_Y'
```

Variable	N	Mean	Median	TrMean	StDev	SEMean
post_x	1000	-1.4813	-1.3014	-1.4399	1.4939	0.0472
post_y	1000	-4.4295	-4.3647	-4.3977	1.3877	0.0439

Variable	Min	Max	Q1	Q3
post_x	-6.5984	2.4086	-2.5086	-0.4160
post_y	-8.9542	0.6882	-5.3661	-3.4154

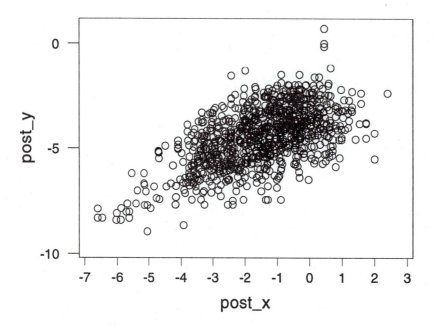

Figure 11.3: Simulated sample from the posterior distribution for contingency table example.

Since U is the log odds-ratio in the 2×2 table, the dotplot and summary statistics are useful for understanding the association in the table. To see whether there is negative association in the table, one can compute the probability that $U < 0$, which is estimated by the proportion of simulated values of U that are negative.

```
MTB > let k1=mean('POST_x'<0)
MTB > print k1
K1        0.837000
```

The probability that the association parameter is negative is .837. Equivalently, the probability that $U > 0$ is .163 — this may not be sufficiently small to reject the statement that the association structure is positive.

11.6   A second example

All of the computational methods illustrated in this chapter make specific assumptions about the shape of the posterior density. In some applications, these assumptions may not apply, and the methods may produce inaccurate results. To demonstrate these problems, suppose that one takes a random sample from a t population with unknown mean M, known scale parameter equal to 1, and $v = 2$ degrees of freedom. If the mean M is assigned a flat prior, and one observes the sample values 2, 3, 10, 12, then the posterior density of M is given by

$$h(M|\text{data}) = \frac{1}{(1 + \frac{(M-2)^2}{2})^{-1.5}(1 + \frac{(M-3)^2}{2})^{-1.5}(1 + \frac{(M-10)^2}{2})^{-1.5}(1 + \frac{(M-12)^2}{2})^{-1.5}}.$$

This posterior density is graphed in Figure 11.4. Note that the density has two modes at approximately 4 and 10. There are two clusters of observations, and the two modes in the density reflect the fact that there is some conflict between the information about the parameter contained in the observations 2, 3 and the information contained in the data 10, 12.

All three computational methods do not work well for this problem. The Laplace algorithm searches for one mode and approximates the density by a normal density about this mode. If one runs 'laplace1', the program will find one of the two modes or the relative minimum of the density depending on the starting guess. Certainly, the use of a normal approximation will be poor. The adaptive quadrature routine does a better job of computing the normalizing constant and indicating the bimodal shape of the posterior density. However, the use of the ten-point grid in 'ad_quad1' is not a fine enough grid for this problem. When the program is run, the computed value of the logarithm of the normalizing constant will oscillate between two distinct values. A finer grid of twenty points would give a more accurate estimate at the mean, standard deviation, and normalizing constant. Last, the Metropolis-Hastings simulation algorithm used in 'metrop' does not perform well due to the two modes. Figure 11.5 gives a histogram of 1000 simulated values produced by the run of 'metrop' starting at the value $M = 5$. Note that, although the algorithm visits both humps of the density, it appears to spend too much time near the second mode $M = 10$ (compare Figures 11.4

and 11.5). Although a relatively large simulation sample has been taken, the algorithm hasn't yet provided a good approximation to the posterior density.

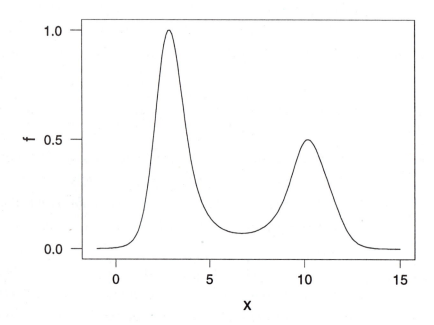

Figure 11.4: Posterior density of mean from a $t(2)$ population.

One also should be cautious about the accuracy of the three algorithms for a two-parameter problem. The Laplace algorithm is inaccurate when the density contains more than one mode. The adaptive quadrature algorithm assumes that both the mean and covariances of the density exist, which may not be the case if the density has a relatively flat tail. The Gibbs algorithm can perform poorly when the two parameters are highly correlated. In this case, for a fixed value of one parameter, the simulated value of the second parameter from the conditional distribution has a limited range, and it can be difficult for the algorithm to move through the entire distribution. Some of these problems will be illustrated in some of the applications discussed in the exercises.

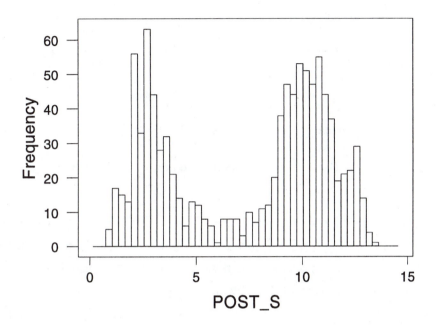

Figure 11.5: Simulated sample from the posterior distribution for t population example.

11.7 Exercises

For each problem, a macro named 'logpost1' or 'logpost2' needs to be written that contains the definition of the logarithm of the posterior density. Once this macro is written, any of the techniques of this chapter can be used to summarize the density.

1. (Estimating an exponential mean) Suppose a random sample is taken from an exponential distribution with mean M. If we reparameterize M to $Y = \log(M)$ and use a noninformative uniform prior on Y, then the posterior density of Y is given by

$$h(Y|\text{data}) = CM^{-n}\exp\{-s/M\},$$

where $M = \exp(Y)$, n is the sample size, s is the sum of the observations, and C is a constant. In exercise 1 of Chapter 9, we illustrate this situation with life-testing data. Five bulbs are tested with observed burn times (in hours) of 751, 594, 1213, 1126, 819. For this data, summarize the posterior density of Y. Estimate the probability that $M > 1000$.

2. (Estimating a normal mean using a Cauchy prior) Suppose that one observation x is normally distributed with mean M and standard deviation 1. If M is assigned a Cauchy prior, then the posterior density for M is given by

$$h(M|\text{data}) = C \exp\left(-\frac{1}{2}(x - M)^2\right) \frac{1}{(1 + M^2)},$$

where C is a constant. For each of the observations $x = 0$ and $x = 5$, summarize the posterior density.

3. (Estimating a truncated normal mean) In exercise 2 of Chapter 10, we consider the problem of estimating a normal mean M when one believes a priori that the mean is positive. Suppose we take a random sample from a normal distribution with mean M and known standard deviation S. Let M have a uniform prior over positive values. If we reparameterize M to $Y = \log(M)$, then the posterior density of Y is given by

$$h(Y|\text{data}) = C \exp\left(-\frac{n}{2S^2}(M - \bar{x})^2\right) M,$$

where $M = \exp\{Y\}$, $\bar{x}$ is the mean of the observations, n is the sample size, and C is a constant. For the data here and $S = 150$, summarize the posterior density of Y.

$$-165, 300, 550, -45, -124, 33, -155, -228, 86, 123, 70, 262, 243, 46$$

4. (Estimating the ratio of two exponential means.) Suppose one observes two independent random samples. The first sample of size m is distributed exponential with mean M_1 and the second sample of size n is distributed exponential with mean M_2. One is interested in estimating the ratio of means $R = M_1/M_2$. If one reparameterizes (M_1, M_2) into $Y_1 = \log R$ and $Y_2 = \log(M_1 M_2)$, then the likelihood function is given by

$$l(Y_1, Y_2) = \frac{1}{M_1^m} \exp\{-s/M_1\} \frac{1}{M_2^n} \exp\{-t/M_2\},$$

where $M_1 = \exp\{(Y_1 + Y_2)/2\}$, $M_2 = \exp\{(Y_2 - Y_1)/2\}$, and s and t are the sum of observations from the first and second samples, respectively. If Y_1 and Y_2 are each assigned uniform priors, then the posterior density of (Y_1, Y_2) is proportional to the likelihood. Exercise 11.10 in Martz and Waller[1] considers the situation where two independent components are tested. The times until failure (in hours) of seven units from the first component are 9.5, 21.0, 40.0, 74.6, 146.3, 148.3, 211.1 and the times until failure of ten units from the second component are 23.1, 29.6, 37.5, 44.3, 44.8, 71.6, 92.2, 101.7, 118.7, 189.9. Assume that the times to failure for units from the two components are exponentially distributed with respective means M_1 and M_2. Find the posterior mean and standard deviation of the logarithm of the ratio of means Y_1. Estimate the posterior probability that $M_2 > M_1$.

5. (Estimating a logistic regression model.) Consider the data in the following table (Mendenhall, et al[2]). The first column represents the number of days of radiotherapy received by each of twenty-four patients. The second column represents the absence (1) or presence (0) of disease at a site three years after treatment. Let p_i denote the probability that the ith patient does not have the disease and x_i the corresponding days of radiotherapy. Suppose that p_i is related to x_i by the logistic model

$$\log\left(\frac{p_i}{1 - p_i}\right) = a + bx_i.$$

The likelihood function of (a, b) is given by

$$l(a, b) = \prod_{i=1}^{24} p_i^{y_i}(1 - p_i)^{1 - y_i},$$

where $y_i = 1$ if the ith patient does not have the disease and $y_i = 0$ otherwise. If one assigns (a, b) a uniform prior, then the posterior density is proportional to the likelihood. Summarize the posterior density of (a, b). Find an approximate 95% interval estimate for b.

---

[1]Martz, H. F. and Waller, R. A. (1982), *Bayesian Reliability Analysis*, New York: Wiley.

[2]Mendenhall, W. M., Parsons, J. T., Stringer, S. P., Cassissi, N. J. and Million, R. R. (1989), "T2 oral tongue carcinoma treated with radiotherapy: analysis of local control and complications," *Radiotherapy and Oncology*, 16, 275-282.

Days	Response	Days	Response
21	1	51	1
24	1	55	1
25	1	25	0
26	1	29	0
28	1	43	0
31	1	44	0
33	1	46	0
34	1	46	0
35	1	51	0
37	1	55	0
43	1	56	0
49	1	58	0

6. (Estimating the median of a Cauchy distribution.) Suppose independent observations $x_1, \ldots, x_n$ are sampled from a Cauchy density with median M and scale parameter 1. If a flat prior is assigned to M, then the posterior density is given by

$$h(M|\text{data}) = \prod_{i=1}^{n} \frac{1}{1 + (x_i - M)^2}.$$

For the data 2, 3, 10, 12, use the three computational algorithms to summarize the posterior density. Compare the three algorithms and comment on their accuracy for this example.

7. (Estimating the location and scale parameters of a t distribution.) Suppose one observes a random sample $x_1, \ldots, x_n$ from a t distribution with location parameter M, scale parameter S, and known degrees of freedom v. If one reparameterizes (M, S) into the real-valued parameters M and $Y = \log(S)$, then the likelihood function is given by

$$l(M, Y) = \prod_{i=1}^{n} \frac{1}{S}\left(1 + \frac{(x_i - M)^2}{S^2 v}\right)^{-(v+1)/2},$$

where $S = \exp(Y)$. Using a constant noninformative prior for (M, Y) and $v = 4$ degrees of freedom, summarize the posterior density for Darwin's data given in exercise 2 of Chapter 6.

8. (Learning about a bivariate normal density) Suppose that one observation (x, y) is taken from a bivariate normal density with means M_x, M_y, scale parameters 1, 1, and known correlation coefficient r. If a flat prior is used and one observes $(x, y) = (0, 0)$, the joint posterior density of (M_x, M_y) is given by

$$h(M_x, M_y | \text{data}) = \exp\{-\frac{1}{2(1 - r^2)}(M_x^2 - 2rM_xM_y + M_y^2)\}.$$

For the correlation value $r = .9$, take a simulated sample from this posterior density using the Gibbs sampler. Using a scatterplot, plot the simulated values contained in the columns 'POST_X' and 'POST_Y'. Repeat this simulation using larger values of the correlation r. Comment on the accuracy of the Gibbs sampler for each simulation run.

List of Minitab Macros

Chapter 2: Simulating Games of Chance

craps: Plays game of craps

To run: Type exec 'craps'. The program will let you play the game a number of times. At the end of the games, some summary statistics are displayed.

yahtzee: Plays simplified version of Yahtzee

To run: Type exec 'yahtzee'. Five dice are rolled three times. After each roll, you decide which dice to keep (if you wish to keep the first, second and fourth dice, you type 1 1 0 1 0). After each roll, the result is printed (two of a kind, small straight, etc.).

yahtz_au: Plays Yahtzee using a particular strategy

To run: Type exec 'yahtz_au'. The computer will decide which dice to keep.

yahtz_re: Plays the computer version of Yahtzee repeatedly

To run: Type exec 'yahtz_re' K, where K is the number of times you wish to repeat the game. (The program 'yahtz_au' must be run first.) The results for each game in coded form are displayed. The results of the first three rolls are placed in the columns 'ROLL_1', 'ROLL_2', 'ROLL_3'. The column 'F_ROLL' contains the results of the final roll.

bball: Simulates a baseball season using a Bradley-Terry model

To run: Type exec 'bball'. One inputs the strength values for the teams. A season is played where each team plays every other team eight games. The results of all of the games are displayed in a two-way contingency table. The numbers of games won by all of the teams are displayed.

bball_re: Repeats 'bball'; simulates a number of baseball seasons

To run: Type exec 'bball_re' K, where K is the number of times you wish to repeat the simulation. (The program 'bball' must be run first.) The numbers of games won for all teams in each season are displayed. The numbers of wins for all teams are stored in the columns 'NWIN_1', 'NWIN_2', etc. ('NWIN_1' contains the number of games won by the team with the first strength value, etc.) The place finishes for all teams are stored in the columns 'PLACE_1', 'PLACE_2', etc.

Chapter 3: Introduction to Inference Using Bayes' Rule

bayes_se: Sets up models, prior probabilities, and likelihoods for Bayes' rule
To run: Type exec 'bayes_se'. You input the number of models, the prior probabilities, the number of observations and names, and the likelihoods.

bayes: Implements Bayes' rule sequentially for independent sequence of outcomes
To run: Type exec 'bayes'. (The program 'bayes_se' must be run first.) You input the sequence of independent observations, and Bayes' rule is implemented sequentially.

Chapter 4: Learning about a Proportion

p_disc: Computes the posterior distribution for p when there is a discrete set of models

To run: Name two Minitab columns 'P' and 'PRIOR'. Place the value of p in 'P' and the prior probabilities in 'PRIOR'. Then type exec 'p_disc'. The posterior probabilities are stored in the column 'POST'.

p_disc_p: Computes predictive distribution for number of successes in future binomial experiment

To run: The values of p are contained in a column 'P' and the probabilities are contained either in the columns 'PRIOR' or 'POST'. Type exec 'p_disc_p'. You will input the number of trials in the future experiment, if you want to use prior or posterior probabilities for p, and the range of number of successes that you are interested in. The predictive distribution is stored in the columns 'SUCC' and 'PRED'.

beta_sel: Finds parameters of beta distribution that matches two predictive probabilities

To run: Type exec 'beta_sel'. You input the probability of a future success and the probability of a second success conditional on a first success. The program gives the matching values of the beta parameters a and b.

p_beta: Summarizes beta distribution for p

To run: Type exec 'p_beta'. After inputting the beta parameters a and b, you can see a graph of the density, compute cumulative probabilities for a list of values of interest, and compute percentiles of the distribution.

p_beta_p: For beta prior or posterior, computes predictive distribution for number of successes in future experiment

To run: Type exec 'p_beta_p'. You input the beta parameters, the number of trials, and the range of numbers of successes that you are interested in. The predictive distribution is stored in the columns 'SUCC' and 'PRED'.

p_beta_t: Tests the hypothesis that p is equal to a specific value

To run: Type exec 'p_beta_t'. You input the value of p that you wish to test, the prior probability of this value, the parameters of the beta distribution under the alternative hypothesis, and the number of successes and failures in the experiment. The output is the Bayes factor and the posterior probability that the proportion is equal to the specific value.

p_hist_p: For a histogram prior for p, finds posterior probabilities of intervals using simulation

To run: Type exec 'p_hist_p'. Suppose that you can divide the interval of plausible p values into equally spaced subintervals and assign a probability to each subinterval. You input the list of midpoints of the subintervals, the corresponding prior probabilities, and the data (number of successes and failures). The posterior probabilities of the intervals are displayed in the output.

Chapter 5: Comparing Two Proportions

pp_disc: Finds posterior distribution for (p_1, p_2) when a uniform prior on a grid is used

To run: Type exec 'pp_disc'. You input the low and high values of each proportion on the grid, the number of models, and the number of successes and failures for each sample. The table of posterior probabilities of (p_1, p_2) is displayed and graphed. The posterior probabilities for the difference $d = p_2 - p_1$ are shown. The values of p_1 and p_2 are stored in the columns 'P1' and 'P2'; the values of the prior and posterior distribution are stored in the columns 'PRIOR' and 'POST'. The probability distribution for d is stored in the columns 'DIFF' and 'P_DIFF'.

pp_disct: Using a prior that gives a specific probability to $p_1 = p_2$, summarizes posterior

To run: Type exec 'pp_disct'. For each proportion, you input the low and high values and the number of models. Also, you input the prior probability that $p_1 = p_2$ and the binomial data. The output is the same as the program 'pp_disc'.

pp_discm: Finds posterior distribution when informative prior is given in table form

To run: Place the values of p_1 in column C1 (say), the values of p_2 in column C2, and the table of prior probabilities in columns C3–CN (each row corresponds to the corresponding value of p_1 and all values of p_2). Type exec 'pp_discm'. You tell the program where the prior is located and input the number of successes and failures for each sample. The output is the same as the program 'pp_disc'.

pp_beta: Using simulation, summarizes distribution of $p_2 - p_1$ when proportions have independent beta distributions

To run: Type exec 'pp_beta'. Input the parameters of the beta distribution for each proportion and the number of values to simulate. Program gives dotplot of distribution of $p_2 - p_1$ and computes probabilities that $p_2 - p_1$ exceeds each of a list of values that is inputted. The simulated samples of p_1, p_2, and $p_2 - p_1$ are stored in the columns 'P1', 'P2', and 'P2-P1'.

pp_bet_t: Constructs test that $p_1 = p_2$ using beta priors

To run: Type exec 'pp_bet_t'. Input the prior probability of equality, the beta distribution for the common value $p_1 = p_2$, the independent beta distributions for p_1 and p_2 when p_1 is not equal to p_2, and the data. The output is the Bayes factor and the posterior probability of equality.

pp_exch: Using an exchangeable prior on the logits, summarizes the posterior distribution of p_1, p_2

To run: Type exec 'pp_exch'. You input the data, the prior standard deviation of the normal prior distribution on the logits, and the number of values to simulate. The simulated samples from the prior and posterior distributions are stored in the columns 'PRIOR_P1','PRIOR_P2', 'POST_P1', and 'POST_P2'.

Chapter 6 - Learning about a Normal Mean

m_disc: Computes the posterior distribution for M when there is a discrete set of models (sampling variance known)
To run: Name two Minitab columns 'M' and 'PRIOR'. Place the value of M in 'M' and the prior probabilities in 'PRIOR'. If data are in raw form, then place it into a particular column of the worksheet . Then type exec 'm_disc'. You input the value of the known standard deviation and the column number of the data. (If data are in summary form, input the sample size, sample mean, and sample standard deviation.) The posterior probabilities are stored in the column 'POST'.

normal_s: Finds parameters of normal distribution that matches two prior percentiles
To run: Type exec 'normal_s'. One inputs two percentiles (probability to the left and the corresponding percentile), and the program gives the mean and standard deviation of the matching normal curve.

m_cont: Summarizes posterior distribution for normal mean M with a normal prior (approximate method)
To run: If data are in raw form, place them into a particular column of the worksheet. (Summary data may also be input.) Type exec 'm_cont'. You input the mean and standard deviation of the normal prior distribution and the column number of the data (if data are in summary form, input the sample size, sample mean, and sample standard deviation). The output is the mean and standard deviation of the approximate normal marginal posterior density for M. Also, the normal parameters of the approximate predictive density for one future observation are given.

normal: Plots and performs normal density calculations
To run: Type exec 'normal'. You input the mean and standard deviation of the normal density. The density is graphed and it computes cumulative probabilities and percentiles of the distribution.

m_norm_t: Constructs test that M is equal to specific value using normal priors (sampling variance known)

To run: Place data into a particular column. (Summary data may also be input.) Type exec 'm_norm_t'. You input the value to be tested, the prior probability of this value, a list of plausible values for the prior standard deviation of M under the alternative hypothesis and the population standard deviation. In addition, you input the number of the column where the data are located. The output is the Bayes factor and the posterior probability of the hypothesis for each value of the prior standard deviation.

m_nchi: Gives exact analysis for M and S using vague priors

To run: Place data into a particular column. (Summary data may also be input.) Type exec 'm_nchi'. Plots and gives summary statistics from the marginal posterior densities of M and the standard deviation S.

Chapter 7: Learning about Two Normal Means

mm_cont: Gives mean and standard deviation for the normal distribution for the difference $M_1 - M_2$ when M_1 and M_2 have independent normal distributions

To run: Type exec 'mm_cont'. You input the mean and standard deviation for each of the two means, and the program gives the mean and standard deviation for the difference in means.

mm_tt: Uses simulation to summarize the difference $M_2 - M_1$ (difference of two t distributions)

To run: Place the two data sets into two columns. (Summary data may also be used.) Type exec 'mm_tt'. The output is dotplots of the marginal posterior densities for M_1 and M_2 and a dotplot and summary statistics for the difference in means.

Chapter 8: Learning about Relationships

lin_reg: Gives inference about slope and predictive response for simple regression model

To run: Place the x and y data in two columns. Type exec 'lin_reg'. You input the numbers of the columns of the data. The output is the mean and standard deviation of the regression slope. In addition, for each value of x that is inputted, the mean and standard deviation of the predicted response are given.

c_table: Computes a Bayes factor for a two-way contingency table using uniform priors

To run: Place the contingency table in consecutive columns of the worksheet. Type exec 'c_table'. Input the numbers of the columns that contain the table. The output is the usual chi-squared test and a Bayes factor against the hypothesis of independence.

Chapter 9: Learning about Discrete Models

mod_disc: Computes a posterior distribution for discrete models

To run: Place model values in the column 'MODEL' and the prior probabilities in 'PRIOR'. Type exec 'mod_disc'. You input the number of the likelihood (binomial, normal, Poisson, etc.) and the data. The posterior probabilities are stored in the column 'POST'.

disc_sum: Summarizes a discrete probability distribution

To run: Type exec 'disc_sum'. Input the numbers of the columns that contain the values of the variable and the probabilities. The distribution is graphed and summaries (mode, mean and standard deviation) are given. Cumulative probabilities and probability intervals are computed for sets of values of interest.

mod_crit: Compares two discrete prior distributions

To run: Place model values in the column 'MODEL' and the two sets of prior probabilities in the columns 'PRIOR1' and 'PRIOR2'. Type exec 'mod_crit'. You input the number of the likelihood (binomial, normal, Poisson, etc.) and the data. The two sets of corresponding posterior probabilities are stored in the columns 'POST1' and 'POST2'. The Bayes factor in support of the first prior is given; this value is stored in the column 'BAYES_F'.

Chapter 10: Learning about Continuous Models

mod_cont: Computes the posterior distribution for continuous models

To run: Simulate a large number of values from the prior distribution of the parameter. Place these values in the column 'PRIOR_S'. Type exec 'mod_cont'. You input the number of the likelihood (binomial, normal, Poisson, etc.) and the data. A simulated sample from the posterior density is stored in the column 'POST_S'.

Chapter 11: Summarizing Posterior Distributions

For one real-valued parameter, place the definition of the logarithm of the posterior density in the macro 'logpost1' — the input column is 'X' and the output column is 'F'.

laplace1: Summarizes a one-parameter posterior density defined in macro 'logpost1' using the Laplace method

To run: Type exec 'laplace1'. Input a guess at the mode and the number of iterations. The output is an estimate at the posterior mode and standard deviation and an estimate at the logarithm of the normalizing constant.

ad_quad1: Summarizes a one-parameter posterior density defined in macro 'logpost1' using adaptive quadrature

To run: Type exec 'ad_quad1'. Input guesses at the mean and standard deviation of the parameter and the number of iterations. Estimates at the posterior mean and standard deviation and the logarithm of the normalizing constant are output. The grid, density values, and weights are stored in the columns 'X', 'F', and 'WT'.

metrop: Summarizes a one-parameter posterior density defined in macro 'logpost1' using simulation

To run: Type exec 'metrop'. Input a starting location and the number of values to simulate. The simulated values are stored in the column 'POST_S'.

For two real-valued parameters, place the definition of the logarithm of the posterior density in the macro 'logpost2' — the input columns are 'X' and 'Y', and the output column is 'F'.

laplace2: Summarizes a two-parameter posterior density defined in macro 'logpost2' using the Laplace method

To run: Type exec 'laplace2'. Input a guess at the mode and the number of iterations. The output is an estimate at the mode and posterior standard deviations and covariance and an estimate at the logarithm of the normalizing constant.

ad_quad2: Summarizes a two-parameter posterior density defined in macro 'logpost2' using adaptive quadrature

To run: Type exec 'ad_quad2'. Input guesses at the posterior moments of the two parameters and the number of iterations. The output is estimates at the posterior moments and an estimate at the logarithm of the normalizing constant. The grid and density values on the grid are stored in the columns 'X', 'Y', and 'F'.

gibbs: Summarizes a two-parameter posterior density defined in macro 'logpost2' using simulation

To run: Type exec 'gibbs'. Input a starting location and the number of values to simulate. The simulated values are stored in the columns 'POST_X' and 'POST_Y'.

Formulas Used in the Macros

Chapter 3: Introduction to Inference Using Bayes' Rule

There are k models $M_1, \ldots, M_k$ with respective initial probabilities $P(M_1), \ldots,$ $P(M_k)$. One observes data D and is given the likelihoods $P(D|M_1), \ldots, P(D|M_k)$. Then the updated probability of model M_i is given by

$$P(M_i|D) = \frac{P(M_i)P(D|M_i)}{\sum_{j=1}^{k} P(M_j)P(D|M_j)}, \ i = 1, \ldots, k.$$

Chapter 4: Learning about a Proportion

1. **Using Discrete Models**

   The model is a proportion p and there are k models $p_1, \ldots, p_k$ with respective prior probabilities $P(p_1), \ldots, P(p_k)$. One observes s successes and f failures. The posterior probability of p_i is given (up to a proportionality constant) by

   $$P(p_i|\text{data}) = P(p_i)p_i^s(1 - p_i)^f.$$

   If the current probabilities of $\{p_i\}$ are represented by $\{P(p_i)\}$, then the predictive probability of s successes in a future sample of size n is given by

   $$P(s) = \sum_{i=1}^{k} \frac{n!}{s!(n - s)!}p_i^s(1 - p_i)^{n-s}P(p_i), s = 0, \ldots, n.$$

2. **Using Continuous Models**

   (a) *Selecting a beta prior:*

   Suppose that one's prior beliefs are represented by a beta(a, b) density. The probability of a future success is r and, if the next observation is a success, the probability of a second success is r^+. Then the values of a and b are

   $$a = \frac{r(1 - r^+)}{r^+ - r}, \ b = \frac{(1 - r)(1 - r^+)}{r^+ - r}.$$

(b) *Inference:*

If p has a beta(a, b) prior, and one observes s successes and f failures, then the posterior density for p is beta$(a + s, b + f)$. If the current density of p is beta(a, b), then the predictive probability of s successes in a future sample of size n is

$$P(s) = \frac{n!}{s!(n-s)!} \frac{B(a+s, b+f)}{B(a,b)}, s = 0, \ldots, n,$$

where $B(a, b)$ is the beta function

$$B(a, b) = \frac{\Gamma(a)\Gamma(b)}{\Gamma(a+b)}.$$

(c) *Testing:*

The hypotheses are H: $p = p_0$, K: $p \neq p_0$. When $p \neq p_0$, alternative values are described by a beta(a, b) density. The data consists of s successes and f failures. The Bayes factor in support of the hypothesis H is given by

$$BF = \frac{P(\text{data}|H)}{P(\text{data}|K)} = \frac{p_0^s (1-p_0)^f B(a,b)}{B(a+s, b+f)}.$$

Chapter 5: Comparing Two Proportions

1. Using Discrete Models

A model consists of the pair of proportion values (p_1, p_2). There is a collection of models $\{(p_{1i}, p_{2j})\}$ with initial probabilities $\{P(p_{1i}, p_{2j})\}$. If the number of successes and failures in the two samples are given by (s_1, f_1) and (s_2, f_2), then the posterior probability of the model (p_{1i}, p_{2j}) is proportional to

$$P(p_{1i}, p_{2j}|\text{data}) = P(p_{1i}, p_{2j}) p_{1i}^{s_1} (1 - p_{1i})^{f_1} p_{2j}^{s_2} (1 - p_{2j})^{f_2}.$$

2. Using Continuous Models

(a) *Inference:*

Before sampling, assume that p_1, p_2 are independent with p_1 distributed beta(a_1, b_1) and p_2 distributed beta(a_2, b_2). If two independent samples are taken with s_1 successes and f_1 failures in the first sample and s_2 successes and f_2 failures in the second sample, then p_1 and p_2 are independent with p_1 distributed beta$(a_1 + s_1, b_1 + f_1)$ and p_2 distributed beta$(a_2 + s_2, b_2 + f_2)$. One can obtain the marginal posterior density of the difference $d = p_2 - p_1$ by simulation. One takes independent samples of p_1 and p_2 values from beta$(a_1 + s_1, b_1 + f_1)$ and beta$(a_2 + s_2, b_2 + f_2)$ distributions. For each simulated pair (p_1, p_2), one computes a value of d. The collection of simulated d values is distributed from the marginal posterior density of interest.

(b) *Testing:*

To test the hypotheses H: $p_1 = p_2$, K: $p_1 \neq p_2$, assume that

   i. under the hypothesis H, the common proportion value $p = p_1 = p_2$ is distributed beta(a, b)

   ii. under the hypothesis K, the proportions p_1 and p_2 are independent with beta(a_1, b_1), beta(a_2, b_2) distributions

Then the Bayes factor in support of the hypothesis H is given by

$$BF = \frac{P(\text{data}|\text{H})}{P(\text{data}|\text{K})} = \frac{B(a + s_1 + s_2, b + f_1 + f_2) B(a_1, b_1) B(a_2, b_2)}{B(a, b) B(a_1 + s_1, b_1 + f_1) B(a_2 + s_2, b_2 + f_2)}.$$

(c) *Using an exchangeable prior:*

   i. If $T_1 = \log(p_1/(1 - p_1))$ and $T_2 = \log(p_2/(1 - p_2))$ are the logits of the proportions, then the belief that p_1 and p_2 are exchangeable is modeled by the two-stage distribution:

      A. conditional on m, T_1, T_2 are independent $N(m, t)$, where $N(m, t)$ denotes the normal distribution with mean m and standard deviation t

      B. m is distributed $N(0, 1)$

   ii. One simulates from this prior distribution by the composition method:

      A. simulate values of m from a $N(0, 1)$ distribution

      B. for each value of m, simulate values of T_1 and T_2 from independent $N(m, t)$ distributions

   iii. One obtains a sample from the posterior density of T_1, T_2 by

      A. if the sample counts are (s_1, f_1, s_2, f_2), compute the likelihood values

$$l(T_1, T_2) = \frac{\exp(s_1 T_1)}{(1 + \exp(T_1))^{s_1 + f_1}} \frac{\exp(s_2 T_2)}{(1 + \exp(T_2))^{s_2 + f_2}}$$

      for all simulated values of T_1 and T_2

      B. take a new simulated sample with replacement from the prior sample of simulated values, with the sampling probabilities proportional to the likelihood values

Chapter 6: Learning about a Normal Mean

1. Sampling Model

Independent measurements are taken from a normal population with mean M and standard deviation h.

2. Using Discrete Models

Assume the value of the standard deviation h is known. A model is a value of the mean M. The possible values for the mean are $M_1, \ldots, M_k$ with respective prior probabilities $P(M_1), \ldots, P(M_k)$. Given a random sample of n measurements with sample mean $\bar{x}$, the posterior probability of M_i is given (up to a proportionality constant) by

$$P(M_i|\text{data}) = P(M_i)\exp(-\frac{n}{2h^2}(M_i - \bar{x})^2).$$

3. Using Continuous Models

(a) *Choosing a prior*

Suppose that two percentiles (p_1, M_1), (p_2, M_2) of the prior density for M are specified (the probability that M is less than M_1 is p_1). The standard deviation and mean of the matching normal prior density are given by

$$h_0 = \frac{M_1 - M_2}{z_1 - z_2}, \quad m_0 = M_1 - h_0 z_1,$$

where z_1 and z_2 are the p_1 and p_2 percentiles of a standard normal curve.

(b) *Approximate method*[1]

i. Initial prior beliefs about the mean M are described by a normal distribution with mean m_0 and standard deviation h_0. The prior precision $c_0 = 1/h_0^2$. If a flat prior is used, $c_0 = 0$. A random sample of size n is taken; the sample mean $\bar{x}$ and the sample standard deviation $s = \sqrt{\frac{\sum(x-\bar{x})^2}{n}}$ are computed.

ii. Estimate the population standard deviation by $h = s(1 + \frac{20}{n^2})$, and let $c = n/h^2$. The posterior density for M is approximately $N(m_1, h_1)$, where

$$m_1 = \frac{c_0}{c_0 + c}m_0 + \frac{c}{c_0 + c}\bar{x}$$

$$h_1 = \frac{1}{\sqrt{c + c_0}}.$$

---

[1] Berry, D. (1996), *Statistics: A Bayesian Perspective*, Belmont, CA.: Duxbury Press, Chapter 11.

iii. The predictive density for a future observation y is approximately normal with mean m_1 and standard deviation $\sqrt{h + \frac{1}{c_1}}$, where $c_1 = c_0 + c$.

(c) *Exact method*

i. Assume that (M, h) have the noninformative prior density proportional to $1/h$.

ii. The marginal posterior density for M is given by the t–form

$$\pi(M|\text{data}) = (S + n(M - \bar{x})^2)^{-\frac{v+1}{2}},$$

where $v = n - 1$ and $S = \sum (x - \bar{x})^2$. Equivalently, the standardized variate

$$\frac{M - \bar{x}}{se},$$

where $se = \sqrt{S/(vn)}$ has a standard t distribution with v degrees of freedom. The distribution of M is denoted $t(\bar{x}, se, v)$.

iii. The marginal posterior density for the variance $V = h^2$ is given by the inverse-chi-squared form

$$\pi(V|\text{data}) = V^{-v/2-1} \exp\{-\frac{1}{2}S/V\}, \; V > 0.$$

The density for the standard deviation h is given by

$$\pi(h|\text{data}) = h^{-v-1} \exp\{-\frac{1}{2}S/h^2\}, \; h > 0.$$

This density is denoted $\chi^{-1}(S, v)$.

(d) *A test of a value of a normal mean*

To test the hypothesis H: $M = M_0$ against the alternative hypothesis K: $M \neq M_0$, suppose that the standard deviation h is known and, under the hypothesis K, M is assigned a normal density with mean M_0 and standard deviation t. Then the Bayes factor in support of the hypothesis H is given by

$$BF = \frac{P(\text{data}|\text{H})}{P(\text{data}|\text{K})} = \frac{\frac{n^{1/2}}{h} \exp\{-\frac{n}{2h^2}(\bar{x} - M_0)^2\}}{(h^2/n + t^2)^{-1/2} \exp\{-\frac{1}{2(h^2/n+t^2)}(\bar{x} - M_0)^2\}}.$$

Chapter 7: Learning about Two Normal Means

1. Sampling Model

Two independent samples are taken. The first sample of size n_1 is selected at random from a normal population with mean M_1 and standard deviation t_1. The second sample of size n_2 is selected from a normal(M_2, t_2) population.

2. Approximate Method

If the posterior density for M_1 can be suitably approximated by a $N(m_1, h_1)$ distribution and M_2 by a $N(m_2, h_2)$ distribution, then the posterior density for the difference in means $d = M_1 - M_2$ is can be approximated by a normal density with mean $m_1 - m_2$ and standard deviation $\sqrt{h_1^2 + h_2^2}$.

3. Exact Method

If the parameters (M_1, M_2, h_1, h_2) are assigned the noninformative density proportional to $\frac{1}{h_1 h_2}$, then M_1 and M_2 have independent t densities of the form

$$\pi(M_1|\text{data}) = (S_1 + n_1(\bar{x}_1 - M_1)^2)^{-\frac{v_1+1}{2}}$$

$$\pi(M_2|\text{data}) = (S_2 + n_2(\bar{x}_2 - M_2)^2)^{-\frac{v_2+1}{2}},$$

where $v_i = n_i - 1$, $\bar{x}_i$ is the sample mean of the ith sample, and $S_i = \sum (x_i - \bar{x}_i)^2$, $i = 1, 2$.

Simulation is used to find the posterior distribution of the difference $d = M_1 - M_2$. A large number of values are simulated from the t marginal density of M_1; the same number of values are simulated independently from the density of M_2. For each pair of simulated values, the difference $M_1 - M_2$ is computed; the resulting simulated sample is distributed from the marginal posterior density of d.

Chapter 8: Learning about Relationships

1. Linear Regression

(a) *The Model*

Observe data $(x_1, y_1), \ldots, (x_n, y_n)$. The values of the x_i are fixed and the response measurements y_i are assumed independent $N(a + bx_i, h)$. The vector of unknown parameters (a, b, h) is assigned a flat noninformative density.

(b) *Inference*[2]

The least-squares estimates of a and b are given by

$$B = r\frac{s_y}{s_x}, \ A = \bar{y} - B\bar{x},$$

where r is the correlation coefficient

$$r = \frac{\overline{xy} - \bar{x}\bar{y}}{s_x s_y}.$$

In this expression, $\bar{x}$ is the mean of the x's, $\bar{y}$ is the mean of the y's, $\overline{xy}$ is the mean of the products $\{x_i y_i\}$, and $s_x = \sqrt{\frac{\sum(x-\bar{x})^2}{n}}$, $s_y = \sqrt{\frac{\sum(y-\bar{y})^2}{n}}$. Estimate the unknown standard deviation by

$$h = (1 + \frac{20}{n^2})s_y\sqrt{1-r^2}.$$

Then the marginal posterior density of the slope b is approximately normal with mean B and standard deviation $\frac{h}{\sqrt{n-2}s_x}$.

(c) *Prediction*[2]

The predictive density of a new observation y at a fixed value of x is approximately normal with mean and standard deviation

$$m_y = A + Bx$$

$$s_y = h\sqrt{\frac{n+1}{n} + \frac{(x-\bar{x})^2}{ns_x^2}}.$$

2. Testing for Independence in a Contingency Table

(a) *The Model*

Observe counts $\{y_{ij}\}$ from a multinomial distribution with sample size n and unknown probabilities $\{p_{ij}\}$. One wishes to test the hypothesis of independence

$$I : p_{ij} = p_{i+}p_{+j},$$

where $\{p_{i+}\}$ and $\{p_{+j}\}$ are the marginal probabilities of the table.

---

[2]Berry, D. (1996), *Statistics: A Bayesian Perspective*, Belmont, CA.: Duxbury Press., Chapter 14.

(b) *The Test*

Under the alternative hypothesis of dependence, assume that the multinomial vector $\{p_{ij}\}$ has a uniform distribution. Under the independence hypothesis, assume that the marginal probability vectors $\{p_{i+}\}$ and $\{p_{+j}\}$ are independent with uniform distributions.

Then the Bayes factor against the hypothesis of independence is given by

$$BF = \frac{D(y+1)D(1_R)D(1_C)}{D(1_{RC})D(y_R+1)D(y_C+1)},$$

where y is the vector of observed counts, y_R is the vector of row totals, y_C is the vector of column totals, 1_R is the vector of ones of length R and $D(v) = \prod \Gamma(v_i)/\Gamma(\sum v_i)$.

Chapters 9 and 10: Learning about Discrete and Continuous Models

1. Discrete Models

(a) *Inference*

Suppose that one takes data from a known sampling density with one unknown parameter M. A specific value for M is called a model. There are k models of interest, where $P(M_i)$ is the prior probability of model M_i. One observes data — $l(M)$ is the likelihood of M that is the probability of observing the data for this particular model. Different choices for the likelihood are listed here. Then the posterior probability of the model M_i is given (up to a proportionality constant) by

$$P(M_i|\text{data}) = P(M_i)l(M_i).$$

(b) *Model criticism*

Suppose for a given set of models $\{M_i\}$, there are two sets of prior probabilities $\{P_1(M_i)\}$ and $\{P_2(M_i)\}$. The posterior probability of model M_i for the jth prior is proportional to

$$P_j(M_i|\text{data}) = P_j(M_i)l(M_i).$$

The Bayes factor in support of the first prior is given by

$$BF = \frac{\sum_{i=1}^{k} P_1(M_i)l(M_i)}{\sum_{i=1}^{k} P_2(M_i)l(M_i)}.$$

2. Inference for Continuous Models[3]

Suppose that the model M is continuous with proper prior density function $\pi(M)$. Let $\{M_i\}$ denote a simulated sample from this prior density. For each simulated value M_i, compute the likelihood $l(M_i)$. A simulated sample from the posterior is found by taking a new sample of values with replacement from $\{M_i\}$, where the probability of selecting a particular value is proportional to its likelihood value.

3. Likelihoods

(a) *Binomial*

One observes a sequence of independent Bernoulli trials with an unknown probability of success M. The observed data is the number of successes s and number of failures f. The likelihood is

$$l(M) = M^s(1-M)^f, 0 < M < 1.$$

(b) *Normal*

One takes a random sample of size n from a normal population with unknown mean M and known standard deviation h. If $\bar{x}$ is the sample mean of the observations, the likelihood is

$$l(M) = \exp\{-\frac{n}{2h^2}(M - \bar{x})^2\}.$$

(c) *Poisson*

Suppose an observation x is taken from a Poisson population with mean tM, where t is the length of the associated time interval, and M is the unknown mean rate in a unit time interval. The likelihood of M is given by

$$l(M) = \exp\{-tM\}M^x, M > 0.$$

(d) *Discrete Uniform*

Independent observations are taken from a distribution uniformly distributed on the integers $\{1, \ldots, M\}$, where the upper limit M is unknown. Let n denote the sample size and m the maximum observation. Then the likelihood of M is given by

$$l(M) = \frac{1}{M^n}, M = m, m+1, m+2, \ldots$$

---

[3]This method of simulating from a posterior density is described in Albert, J. (1993), "Teaching Bayesian statistics using sampling methods and Minitab," *The American Statistician*, 46, 167-174.

(e) *Hypergeometric*

Suppose a finite population of a known size N consists of successes and failures, where M is the unknown number of successes. A sample of size n is taken without replacement from the population, and s successes in the sample are observed. The likelihood of M is

$$l(M) = \binom{M}{s}\binom{N-M}{n-s}, \ s \leq M \leq N - n + s.$$

In the computation of the binomial coefficients, Stirling's approximation is used:

$$x! \approx \sqrt{2\pi}\exp(-x)x^{x+.5}$$

(f) *Capture-Recapture*

To learn about the size M of a finite population, a capture-recapture sampling scheme is used. A known number K of marked items are introduced into the population, and a sample of size n is selected without replacement from the new population, obtaining x marked items. The likelihood function of M is given by

$$l(M) = \frac{\binom{M}{n-x}}{\binom{M+K}{n}}, \ M \geq n - x.$$

In the computation of the binomial coefficients, Stirling's approximation is used.

(g) *Exponential*

A random sample of size n is taken from an exponential population with mean M. If the sum of the observations is denoted by s, the likelihood is given by

$$l(M) = M^{-n}\exp\{-s/M\}, \ M > 0.$$

Chapter 11: Summarizing Posterior Distributions

Let M denote the vector of parameters of interest. Let $\pi(M)$ denote the prior density and $l(M)$ the likelihood function. The posterior density is given by

$$\pi(M|\text{data}) = H^{-1}\pi(M)l(M).$$

The logarithm of the posterior density will be denoted by $f(M) = \log \pi(M|\text{data})$. The normalizing constant of the posterior density is given by

$$H = \int \pi(M)l(M)dM.$$

1. The Laplace Method[4]

Let M_i be a current guess at the mode of the posterior density. Then, by the Newton-Raphson algorithm, the updated estimate at the mode is given by

$$M_{i+1} = M_i - f_2^{-1}(M_i)f_1(M_i),$$

where $f_1(M_i)$ and $f_2(M_i)$ are the gradient and hessian, respectively, of the log posterior density f evaluated at the current estimate of the mode. The algorithm is continued until convergence; let $\tilde{M}$ denote the posterior mode. The estimate at the posterior variance-covariance matrix is given by $V = -f_2^{-1}(\tilde{M})$. Approximately, the posterior density is $N(\tilde{M}, V)$. The estimate at the normalizing constant is given by

$$H \approx (2\pi|V|)^{p/2}\pi(\tilde{M})l(\tilde{M}),$$

where p is the number of parameters and $|V|$ denotes the determinant of the matrix V.

2. Adaptive Quadrature[5]

For a single parameter, the adaptive quadrature method is based on the Gauss-Hermite rule

$$H = \int \pi(M)l(M)dM \approx \sum_{i=1}^{k} w_i\pi(M_i)l(M_i),$$

where $w_i = h_i\sigma\sqrt{\pi}$, $M_i = \mu + t_i\sigma\sqrt{\pi}$, and $\{t_i\}$ and $\{h_i\}$ are zeros and weights, respectively, of the Hermite polynomial $H_k(t)$. In the program 'ad_quad1', a $k = 10$ point grid is used. To implement this rule, one guesses at the posterior mean μ and posterior standard deviation σ at the posterior density. These moment guesses determine the grid $\{M_i\}$ and weights $\{w_i\}$ of the rule. The algorithm estimates H and new values of μ and σ. The algorithm is iterated until the estimates at H and the moments stabilize. The resulting summary is that M is approximately distributed according to a discrete distribution with mass points $\{M_i\}$ and probabilities proportional to $\{w_i\pi(M_i)l(M_i)\}$.

A similar algorithm is used to compute the normalizing constant for a two-parameter posterior density on a product grid. To increase the efficiency of

---

[4] An introduction to the use of the Laplace method is contained in Chapter 3 of Tanner, M. A. (1993), *Tools for Statistical Inference: Methods for the Exploration of Posterior Distributions and Likelihood Functions*, second edition, New York: Springer-Verlag.

[5] This procedure is described in Naylor, J. C. and Smith, A. F. M. (1982), "Applications of a method for the efficient computation of posterior distributions," *Applied Statistics*, 31, 214-235.

the algorithm, one first performs a linear transformation of the form $\gamma_1 = M_1, \gamma_2 = M_1 + aM_2$ such that the correlation between the transformed parameters γ_1 and γ_2 is equal to 0. A product grid rule is used on the space of the transformed parameters.

3. Simulation

(a) *Metropolis-Hastings*[6]

For a single parameter M, let $M^{(i-1)}$ denote the current simulated value. A candidate value M^C is simulated from a normal density with mean $M^{(i-1)}$ and standard deviation c, where c is a known scale factor. One computes the probability

$$p = \min\{\frac{\pi(M^C)l(M^C)}{\pi(M^{(i-1)})l(M^{(i-1)})}, 1\},$$

and simulates a random uniform variate U. Then the next simulated value, $M^{(i)}$, is equal to M^C if $U < p$, and is equal to $M^{(i-1)}$ otherwise. Starting at the value $M = M^{(0)}$, the stream of simulated values $\{M^{(1)}, M^{(2)}, \ldots\}$ is approximately distributed according to the posterior density.

(b) *Gibbs sampling*[7]

For a two-parameter problem, let $(M_1^{(i-1)}, M_2^{(i-1)})$ denote the current simulated value of (M_1, M_2). A new simulated pair of values is obtained by sampling from two conditional posterior distributions. First, the current value of M_1 is fixed and a value of M_2 is simulated from the conditional posterior density of M_2 conditional on $M_1 = M_1^{(i-1)}$. This one-parameter simulation is accomplished by the Metropolis-Hastings algorithm. Call the simulated value $M_2^{(i)}$. Then a value of the first parameter M_1 is sampled from the distribution of M_1 conditional on $M_2 = M_2^{(i)}$ — this value is denoted by $M_2^{(i)}$. If the process starts at the point $(M_1^{(0)}, M_2^{(0)})$, then the simulated stream of values $\{(M_1^{(1)}, M_2^{(1)}), (M_1^{(2)}, M_2^{(2)}), \ldots\}$ is approximately a sample from the joint posterior density.

---

[6] An introduction to the Metropolis-Hastings algorithm is contained in Chib, S. and Greenberg, E. (1995), "Understanding the Metropolis-Hastings Algorithm," *The American Statistician*, 49, 327-335.

[7] An introduction to the Gibbs sampler is contained in Casella, G. and George, E. (1992), "Explaining the Gibbs sampler," *The American Statistician*, 46, 167-174.

Index